Science, Technology, and Canadian History

Les Sciences, la technologie et l'histoire canadienne

Science, Technology, and Canadian History

Les Sciences, la technologie et l'histoire canadienne

The First Conference on the Study of the History of Canadian Science and Technology

Premier Congrès sur l'histoire des sciences et de la technologie canadiennes

Kingston, Ontario

Edited by
Richard A. Jarrell and Norman R. Ball

Wilfrid Laurier University Press

Canadian Cataloguing in Publication Data

Main entry under title:

Science, technology, and Canadian history = Les
Sciences, la technologie et l'histoire canadienne

Text in English or French.
Record of the papers and workshops of the first
National Conference on the History of Canadian
Science and Technology, 1978.
Bibliography: p.
Includes index.
ISBN 0-88920-086-6 pa.

1. Science — Canada — History — Congresses.
2. Technology — Canada — History — Congresses.
I. Jarrell, Richard A., 1946-   II. Ball, Norman R.,
1944-   III. Title: Les Sciences, la technologie
et l'histoire canadienne.

Q127.C3S44   509.71   C80-094306-6E

Données de catalogage avant publication (Canada)

Vedette principale au titre:

Science, technology, and Canadian history = Les
Sciences, la technologie et l'histoire
canadienne

Texte en anglais ou en français.
Compte-rendu des ateliers et rapport des exposés
présentées au premier Congrès sur l'histoire
des sciences et de la technologie canadiennes.
Bibliographie: p.
Comprend un index.
ISBN 0-88920-086-6 pa.

1. Science — Canada — Histoire — Congrès.
2. Technologie — Canada — Histoire — Congrès.
I. Jarrell, Richard A., 1946-   II. Ball, Norman R.,
1944-   III. Titre: Les Sciences, la technologie
et l'histoire canadienne.

Q127.C3S44   509.71   C80-094306-6F

Copyright © 1980
WILFRID LAURIER UNIVERSITY PRESS
Waterloo, Ontario, Canada   N2L 3C5
80 81 82 83 4 3 2 1

31,177

Cover design by David Antscherl

TABLE OF CONTENTS/TABLE DES MATIERES

FOREWORD
Bruce Sinclair

About twenty years ago, Professor A. Hunter Dupree
wrote an article for the *American Historical Review* entitled "The History of American Science: A Field Finds Itself." As we reflect upon the proceedings of this conference, in search of the history of Canadian science and
technology, we may take some comfort and instruction from
his essay.

To a greater extent than did Dupree, we have first to
overcome a general tendency to disbelieve there even is
such a thing as a history of Canadian science or technology.
Not only, after all, are there no Bessemers, Pasteurs, or
Darwins in our past--there are no Henry Fords or Thomas
Edisons. Apart from the discovery of insulin, about which
the most interesting thing to know might be how it happened
here instead of the half dozen other places in the world at
which parallel work was simultaneously underway, and apart
from the less than conclusive claim that Alexander Graham
Bell invented the telephone here, there are no stirring discoveries apparently due to peculiarly Canadian factors.

And, as if this absence of heroes were not enough, we
must also struggle against the long-cherished image of scientists that they deal in an international currency, that
technical formulas do not reflect political entities, or
that something called Canadian chemistry is a contradiction
in terms.

Finally, in our efforts to pursue the history of Canadian science and technology, we have to battle the notion
that anything done here was, in any event, a pale imitation
of more creative work done somewhere else. That self-dep-
recatory sense of our past is deeply ingrained in the
national character, but its roots lie also in the already
mentioned hero approach to history--that somehow only
"original" discovery counts, even though the historical re-

Bruce Sinclair

Il y a une vingtaine d'années, le professeur A. Hunter
Dupree publiait un article dans l'*American Historical
Review*, article intitulé "The History of American Science:
a Field Finds itself." Maintenant que le moment est venu
d'émettre quelques réflexions sur les actes de ce congrès
et que nous partons en quête de l'histoire des sciences et
des techniques au Canada, nous pouvons chercher encourage-
ments et conseils dans cet essai.

Plus encore que Dupree, nous devons nous inscrire en
faux contre le préjugé général selon lequel il n'y aurait
ni histoire des sciences, ni histoire des techniques
spécifiques au Canada. Il faut bien avouer que l'on ne
trouve aucun Bessemer, aucun Pasteur, aucun Darwin dans
notre passé, pas plus que de Ford ou de Thomas Edison. Il
y a bien la découverte de l'insuline; mais ce qui est le
plus susceptible d'intéresser l'historien en l'occurrence,
c'est peut-être pourquoi cette découverte eut lieu au Canada plutôt que dans une demi-douzaine d'autres endroits où
des recherches parallèles étaient menées au même moment.
Il y a bien la prétention plus ou moins bien fondée
qu'Alexander G. Bell a inventé le téléphone au Canada;
mais, ormis ces deux événements, aucune découverte glori-
euses, à ce qu'il semblerait, n'a été le produit de
facteurs appartenant en propre au Canada.

Non seulement les héros manquent à l'appel, mais, de
plus, nous devons lutter contre la représentation, privi-
légiée depuis longtemps, du scientifique, cet homme qui
mène ses transactions en "devises" internationales; contre
l'idée aussi que le technique ne reflète en aucune manière
la politique, et que l'énoncé "chimie canadienne" est con-
tradictoire. Finalement, en tenant l'étude de l'histoire
des sciences et des techniques canadiennes, nous devons
également rejeter l'idée que, de toute façon, ce qui se

cord usually demonstrates the multiplicity and simultaneity of invention.

Yet, surely it is clear that there is a history of Canadian technology and science. There have been practitioners of the varied arts and skills those words encompass, there has been vast agricultural development and extensive exploitation of natural resources, there have been institutions that have fostered research and teaching, and there have been all the other activities that involve science and technology in any economically and industrially developed country. Just as surely, our search for the history of Canadian science and technology must go beyond efforts to identify long-neglected people who did not get the credit they deserve. We must go beyond the description simply of what happened to engage the more challenging historical questions of why and how it happened. Most important, I believe, we must recognize that science and technology are as much the products of a culture as art or literature.

Indeed, to some degree we already know that truth. From the outset, for example, the Engineering Institute of Canada--originally known as the Canadian Society of Civil Engineers--took a position on the licensing of professional engineers quite unlike the stance of similar societies in Great Britain or the United States and for reasons that were directly related to the question of national identity. We know, too, that Continental models of science education were taken up more slowly in Canada than they were in the United States, but more rapidly than in Britain, just as we know there were differences in science education between English- and French-speaking Canada, even if they have yet to be systematically analyzed. And in much the same vein, we know that public health institutions developed unique forms in Canada. In other words, it is already apparent that if one seeks to identify the elements of Canadian culture, important clues are to be found in the institutions of science and technology.

I think there is also evidence to suggest that the content of science and technology in this country reflects the national condition, too. Consider, for example, the

faisait ici ne constituait qu'une piètre imitation de
travaux originaux provenant d'un autre pays. Cette dé-
valorisation de notre passé est profondément inscrite dans
le tempérament national, mais elle est aussi due à l'importance que l'on accorde à l'histoire des grands hommes,
cette histoire selon laquelle seules les inventions "ori-
ginales" comptent, et ce en dépit du fait que les études
historiques fouillées montrent souvent que l'invention,
loin d'être ponctuelle, est constituée d'événements mul-
tiples et simultanés.

Cela dit, il faut se rendre à l'évidence: l'histoire
des sciences et des techniques canadiennes existe. Ces
mots qui recouvrent de nombreux arts et métiers, recouvrent
également les hommes qui les pratiquent. L'agriculture
s'est énormément développée; les ressources naturelles sont
exploitées sur une vaste échelle; des institutions ont
rendu possibles la recherche et l'enseignement; enfin on
retrouve au Canada toutes les activités d'un pays indus-
trialisé, qui incorporent les sciences et les techniques.
Il faut également se rendre à l'évidence que, dans notre
quête vers l'histoire des sciences et techniques canadiennes, nous ne devons pas nous limiter à l'identification
d'individus oubliés depuis belle lurette dans le but de
leur accorder les honneurs qu'ils méritent. Nous devons dé-
passer la simple description pour affronter le comment et
le pourquoi de ce qui s'est passé, tâche autrement difficile. A mon avis, la chose la plus importante à bien saisir,
c'est que les sciences et les techniques sont tout autant
les produits d'une culture que l'art ou la littérature.

D'ailleurs, dans une certaine mesure, nous reconnais-
sons la vérité de cet énoncé. Dès ses origines, par exemple,
l'*Institute of Canadian Engineers*, d'abord connu sous le
nom de *Canadian Society of Civil Engineers*, adopta à l'égard
de la licence professionnelle des ingénieurs une attitude
tout à fait différente de celle d'organismes similaires en
Grande-Bretagne ou aux Etats-Unis, et ceci pour des raisons
directement liées à des questions d'identité nationale.
Nous savons aussi que les modèles européens en matière
d'éducation scientifique furent adoptés plus lentement au

observation of John Anderson, an acute student of machine
tool design and British commissioner to the American centennial exhibition in Philadelphia in 1876, who claimed
in his official report that Canadian machine tools had a
character "all their own." I expect you could yourselves
multiply the examples of similar observations.

Indeed, my own conviction is that we shall discover
science and technology to be close to the centre of Canadian experience, that even if unexplored there are inti-
mate connections between science and its applications, on
the one hand, and, on the other hand, long and closely held
Canadian political, economic, and social ambitions.

At the least, then, our tasks at this conference are
to identify the resources and problems of this challenging
new field of study and to know ourselves as coworkers,
that we may better understand the important work that lies
ahead of us. The country does not have a coherent science
policy, I suspect, because it is unconscious of its science
history. Our job is to demonstrate there is one, full of
richness and complexity, and that knowledge of it will better inform the present just as it better illuminates the
past.

Canada qu'aux Etats-Unis, mais plus rapidement qu'en
Grande Bretagne, comme nous savons aussi que l'enseignement des sciences diffère entre les anglophones et les
francophones du Canada, même si ces différences n'ont pas
été étudiées systématiquement. Dans le même ordre d'idées,
nous savons que les institutions de santé publique se sont
développées d'une manière très particulière au Canada.
Bref, il est d'ores et déjà acquis que, pour identifier les
éléments scientifiques d'une culture canadienne, de préci-
eux indices pourront être mis à jour en scrutant les institutions où s'abritent les sciences et les techniques.

A mon avis, certains faits permettent de suggérer que
le contenu des sciences et des techniques au Canada reflète
aussi la situation nationale. On peut s'appuyer, par exemple, sur l'observation de John Anderson, habile spécial-
iste des machines-outils et délégué britannique pour l'ex-
position du Centenaire des Etats-Unis tenue à Philadelphie
en 1876. Celui-ci prétendait dans son rapport officiel
que les machines-outils canadiennes possédaient un style
"très particulier." J'imagine que beaucoup d'entre nous
pourrions multiplier des examples ou des observations de ce
genre.

De fait, et j'en suis convaincu, nous découvrirons que
les sciences et les techniques se situent près du centre de
l'expérience collective canadienne. Sans même les avoir encore étudiées, on peut avancer qu'il existe des rapports
étroits entre les sciences et leurs applications, d'une
part, et, d'autre part, entre les sciences et les ambitions
politiques, économiques et sociales les plus importantes
pour le Canada.

Dans ce congrès, notre tâche minimale aura été d'identifier les ressources documentaires disponibles et les
problèmes propres à ce nouveau domaine d'études qui nous
pose un défi. Ce congrès nous aura aussi permis de faire
connaissance avec nos collègues de façon à mieux saisir la
nature du travail important qui reste à faire. A cet égard,
j'ai l'impression que si le pays n'a pas de politique cohé-
rente dans le domaine de l'activité scientifique, c'est
parce qu'il n'est pas conscient de son histoire scienti-

fique. A nous de montrer que cette histoire existe, à la fois riche et complexe, et que la connaître permettra au présent d'être mieux informé tout en offrant une vision plus claire du passé.

ACKNOWLEDGMENTS AND ORGANIZERS/
REMERCIEMENTS ET COMITE ORGANISATEUR DU CONGRES

Organizing Committee/Comité organisateur

Dr. A. W. Tickner (Chairman/Président)

Mr. N. R. Ball

Prof. C. E. S. Franks

Prof. J.-C. Guédon

Prof. J. B. Sinclair

The Organizing Committee wishes to thank the following
institutions and organizations for their contributions and
financial support of the conference/Le Comité organisateur
voudrait remercier les institutions et les associations
suivantes de leur aide ainsi que leurs contributions finan-
cières au congrès.

Queen's University

Social Science and Humanities Research Council of Canada/
    Conseil de recherches en sciences humaines

National Research Council of Canada/Conseil national de re-
    cherches du Canada

Science Council of Canada/Conseil des sciences du Canada

National Museums of Canada/Musées nationaux du Canada

Hannah Institute for History of Medicine

Atomic Energy of Canada, Ltd./Energie atomique du Canada,
    Ltée.

Aluminum Company of Canada, Ltd.

The Editors wish to thank Prof J.-C. Guédon for translation
and Mrs. Martha Jarrell for typing the manuscript/Les ré-
dacteurs remercient Prof J.-C. Guédon de ses traductions,
et Mme. Martha Jarrell de la dactylographie.

INTRODUCTION

THE STUDY OF THE HISTORY OF CANADIAN
SCIENCE AND TECHNOLOGY
Richard Jarrell & Norman Ball

This volume, although the first of its kind ever
produced in this country, marks the end of an era, the pre-
history of a professional history of Canadian science and
technology. The Kingston Conference was not the first
meeting of professional historians to debate the issues
of an emerging field, but it was the culmination of such
meetings, it was the best of such meetings, and, we may
hope, it was the last of such meetings. Some 150 scholars,
students, archivists, librarians, and governmental figures
exchanged information, argued, planned, and generated an
intensive excitement the like of which the writers have
never before had the pleasure of witnessing. Despite a
stunning diversity of interests and backgrounds, there
emerged a rare unanimity: that the field of the history
of Canadian science and technology has come of age and
must be pursued on all fronts. Should this field fail to
bloom, it will not be from the lack of enthusiasm and
commitment exhibited at Kingston.

A glance over the contents will show the reader the
sheer breadth of this field, a field no doubt destined to
develop into a number of interrelated disciplines, but
which is now held together by the requirements of a small
group facing common problems. This collection represents
the state of the art and, indeed, the state of the hopes
and aspirations of its participants. And, although the
reader will see the formal record of this meeting, what
is missing (and what can never be adequately recorded) is
the table talk, the corridor conversations, the excitement
of the debates over glasses of beer, in short, the collective charging of intellectual batteries. This volume
records not history but how to encourage the writing of
history. The talking is now done and the writing of

1

history continues, in growing volume, for the field has found itself.

Historical writing on Canadian science and technology (and we include the history of medicine) is not new. Examples from the nineteenth century abound, but what is of recent origin is the interest in this subject held by professional university historians. One can date this awakening for the history of Canadian science precisely: June 1972 at the Montreal meeting of the Canadian Historical Association (CHA). At that meeting, there was a session of the history of Canadian science, with two papers read by graduate students of the Institute for the History and Philosophy of Science and Technology at the University of Toronto. At the same time, students of Canadian science were beginning to enter the programme of Canada's other graduate institution, the Institut d'histoire et sociopolitique des sciences at the Université de Montréal. Several students of both schools have since become prominent in the field. While the CHA never repeated this experiment, annual sessions on Canadian science and technology became a feature of the meetings of the Canadian Society for History and Philosophy of Science (CSHPS) from 1973 onward. At its Laval meeting in the summer of 1976, a group of historians and archivists met informally to discuss the lack of organization of what was clearly an emerging discipline. Two concrete proposals came from that meeting. The first was a meeting of historians organized by Donald Phillipson, then with the oral history project of the NRC, along with A. W. Tickner, the NRC archival officer, and held in Ottawa in April, 1977. Attended by more than twenty persons, the meeting debated a number of issues and reviewed common problems. The second outcome of the Laval meeting was a national newsletter for historians of Canadian science and technology, the HSTC Bulletin, edited by the writers. The first issue appeared in November 1976 and this bulletin is now in its third volume with a rapidly growing readership.

A second Ottawa meeting on archival issues was

organized by Phillipson, N. R. Ball, and Tickner in
September, 1977. It was at this meeting that the idea of
a national conference was broached and a committee composed
of Ball, C. E. S. Franks, J.-C. Guédon, J. B. Sinclair,
and Tickner began planning for it. A year's work culmina-
ted in the Kingston conference in November 1978.

The hallmarks of a new discipline are normally the
formation of a society and the institution of a journal,
but neither of these has come about for this field. These
possibilities were weighed at the Laval meeting and its
successors, but the consensus has been that the manpower
is too thin, too scattered, and too diverse to call a
society into being. An attempt to form a semi-permanent
working group within the CSHPS for the history of Canadian
science and technology in 1975 was initially encouraged
but then repudiated. While there is now an *entente
cordiale* between the society and its members in this field,
it is unlikely that CSHPS will be the vehicle for a group
as diverse as this. In fact, no existing society can meet
such requirements. Whether an independent organization
is necessary remains to be seen, but up to this point, the
informal approach has worked admirably for the growing
network of scholars. The possibility of a journal devoted
to the field has been championed by Phillipson since 1976,
but the--so far--small volume of excellent work and
the increasingly difficult financial picture for academic
journals seems to doom that project for the immediate
future. For the moment, the *HSTC Bulletin* partially ful-
fills the needs of the field.

There is no doubt that a journal lends a high profile
to a field and that it can stimulate research. At this
point in the history of the study of Canadian science and
technology, however, there does not seem to be the
quantity of high-class material necessary to sustain a
regular respectable journal. But all of this reflects the
situation in the academic community, which is perhaps some
twenty years behind its American counterpart.

If we look beyond the academy, we do find considerable
activity much of which, until the reports in this volume

appear, is largely unknown. Indeed, the bulk of historical writing in this field is produced by historians -- both professional and amateur -- in scientific departments of government (e.g., the recent National Research Council divisional histories, various government departmental histories, etc.), from museums (especially the National Museum of Man in Ottawa), and from government historical reconstruction agencies. In most cases, the type of research being published by these writers is much narrower in focus than one might expect from a mature academic community, but they are providing the grist for academic mills.

Both the papers of these proceedings and the discussions they represent clearly show a large amount of interest in the history of Canadian science and technology. We should not underestimate this well of enthusiasm. It is not true to say that there is little interest; it may be closer to the truth to say that this interest is not coordinated in any reasonable fashion. Those in a position to make decisions that could distinctly aid the field-- for example, museum directors, academic administrators, educational consultants, archivists, and a variety of governmental officials--simply do not realize the breadth and depth of this interest. It is not their fault if they ignore something no one communicates to them. Therefore, if there is one overriding desideratum for this field, it is the need for better coordination.

The reader cannot but notice that one theme that runs through many of the papers is the lack of money available for suitable historical research and publication. We do not believe this is true. Money is available, but few ask for it. There are many private foundations which, if properly apprised of the value of the history of Canadian science and technology, would make considerable funds available. Likewise, there has been a great deal of discussion, both at the Kingston Conference and its two Ottawa predecessors, about the neglect of archivists in making accessible scientific and technological materials. We again take issue with this. If there is no obvious

demand for a particular kind of material, no overworked
archivist is going to go out of the way to procure it, nor
will his superiors make money available no matter how
ardent the plea.

The varieties of historical writing on Canadian
science are too few and the total literature too small
to differentiate many obvious historiographical viewpoints.
Indeed, debate on theoretical issues is rare in Canadian
history in general; some of the more heated are in social
history, but that field is young. In the history of
Canadian science, writers are not yet conscious of major
differences or, if they are, they refrain from discussing
them. The first category one can identify is of those
earlier (and a few contemporary) survey works that do
little more than enumerate names and events. These are
not all even heroic histories but chronicles. H. M. Tory's
History of Science in Canada may serve as the archetype of
this genre. More advanced examples of this early work
which draw attention to issues, but in unexplicit ways,
are exemplified by the works of Léon Lortie and Jacques
Rousseau. Lortie, in particular, advances the notion that
French Canada did pursue science in the nineteenth century.
This is a form of revisionism, a reaction to the more
traditional Quebec picture of francophones as cultural
rather than scientific intellectuals. The present generation of writers is emerging and redefines Lortie's view
from a more self-consciously theoretical position. Raymond
Duchesne's recent work falls into this category; while he
agrees with Lortie that French Canadians "did science," he
argues that the science they did resembled that of no one
else. His paper in this collection carries on in this
vein.

English-Canadian writing has developed little more
than this. One of the few position statements is in
T. H. Levere and R. A. Jarrell's *A Curious Field-book*,
which takes the view that Canadian science is an unique
creation arising from the European scientific traditions
confronting the realities of the Canadian situation. One

might almost call this view a "dialectical frontierism"; compared with Canadian political historical writing, it comes closer to the views of Arthur Lower than to those of, say, Donald Creighton. Levere's article in this collection develops this theme, and both he and Jarrell would stress this dynamic tension while admitting that their view is oversimplified.

Most historical writing on Canadian science has been institutional history and in the majority of such works the contexts are so narrow as to provide little help in forming more general historiographical concepts. In a sense, however, this is not unexpected. Canadian political or constitutional history, for example, has such rich data that theorizing comes easily; in the history of Canadian science, so little is known in detail that theory can act only heuristically, not as explanation, although these should grow together as the data accumulate. Institutional and biographical writing have been the meagre staple so far, but there are a few signs that social history of science is beginning to awaken.

The pattern of historical writing on Canadian engineering and technology is similar to that for science. The most serious lack is that of both a workable interpretive framework and a well-developed body of analytical articles, however scattered. The disparate body of biographies in the heroic mould and institutional histories which are heavily genealogical and chronological appear to have been stillborn. J. J. Brown's *Ideas in Exile* is one of the only attempts to provide an analytical framework but its thesis is oversimplified and untenable. Brown's view is that Canadians think up the ideas first and others capitalize on them and get the credit. The failure of this thesis to provoke extensive rebuttal or further research raises the question of why Canadian historians refused to take technology and engineering as a subject of serious sustained historical inquiry. It is somewhat curious that a nation founded on fish, fur, lumber, wheat, and minerals, a nation heavily dependent on various resource exploitation technologies and glued together by a force-fed railway

should find so little place for historical enquiry into
foundations rather than superstructure. The neglect
cannot be attributed to a lack of printed indicators. A
glance at works such as G. F. Henderson, *Federal Royal
Commissions in Canada, 1867-1966*, reveals a long tradition
of recognizing that technology is a formative influence in
Canadian life. The work of Innis and Lower documents the
importance of technology. Innis in particular had a personal interest in the working and development of machinery
but neither he nor his disciples made this the subject of
historical inquiry. The resultant loss of one dimension in
Canadian culture is a serious one which cannot be explained
readily. The degree to which its absence has been accepted
as part of professional historical inquiry can be seen in
university calendars, museums, archives, and analytical
writings such as Carl Berger, *The Writing of Canadian
History*, which barely mentions science or engineering.

While one could continue documenting the chronicles
of missed opportunities the important point is that the
pattern is changing. Without the prodding of outside
inquiry, cultural and historical agencies are beginning
to look at neglected areas. One can only speculate on how
serious inquiry might accelerate this process of change.
The brightest spot on the horizon is the work being done
by graduate students. The subjects of a growing number of
theses and dissertations indicate a healthy growth of research in areas related to technology and engineering.
Students trained and influenced by historians who identified significant elements of technology and engineering
but did not explore them are building upon this legacy.
While retrospective reassessment is one of the signs of
the maturing of Canadian historical studies, it should not
be allowed to overshadow the new growth in Canadian historical studies which should greatly enrich Canada's self-
image and understanding. This widening of research
interests is part of a normal pattern of growth which other
countries have or will participate in. The recognition
that the Canadian pattern has not been abnormal makes the
most important question where we go from here rather than
why it took so long.

THE KINGSTON CONFERENCE AND BEYOND

C. E. S. Franks

When I was talking with one of the editors of this
volume about what might go into an introduction, he suggested I should write about an "outsider's view" of the
conference. This took me slightly by surprise, for up to
then I hadn't considered myself an outsider: I had been
in on planning from the beginning, had been secretary of
the conference organizing committee, had handled fund-
raising, and had done much of the arranging of the programme. But the more I thought about it, the more I
realized I was an outsider. There are two institutes in
Canada for the study of the history of science and technology, and a few small departments or near-departments.
I don't belong to any of them, and the discipline which I
am associated with, political science, has not distinguished itself for its interest in the role of science
and technology in Canadian political development.

However, I began to consider further that if I was an
outsider, there must be insiders, and if this were so they
must, if anywhere, have been represented at our conference,
for which enrolment was open, and expenses paid for all
needy registrants. But in looking over the list of registrants I was struck by two things.

First, they came not only from all parts of the
country (and overseas), but also from almost every part of
Canadian intellectual life. There were professors of
geography, political science, business, economics, and
history; professors of the history of medicine and science;
professors of engineering and natural science; representatives of museums, archives, and funding agencies; deans of
engineering and arts; businessmen, scientists, administrators; librarians and bibliographers; retired scientists

and engineers; professional journalists and writers; and
many others. In no way could the group be identified as
a "school" or "discipline" in the traditional university
sense. Their present occupations were as varied as their
academic backgrounds. What did unite them was a common
interest in the Canadian heritage in science and technology, and a common concern that this heritage not be
lost, but preserved, better understood, and better known.

Secondly, some groups were noticeable by their absence. The lack of participation by the National Museum
of Science and Technology and the reasons for it have been
commented on in several of the sessions and need no further
attention here; what was even more striking to me was the
small number of representatives of Canadian professional
historians. No historians from Queen's University, with
the exception of one associate dean of graduate studies,
came to the conference, even though Queen's was the host.
It is possible to entertain more than a suspicion that this
virtual boycott by professors of history came about because
of the small place that the history of science and technology plays in our departments of history, particularly
when the science and technology concerned are Canadian.
History, like all other academic disciplines, has its
conventions regarding what is good scholarship and what is
not, and so far the study of the history of Canadian
science and technology is at the fringes, hardly respectable.

My conclusion is that the persons interested in the
study of the history of Canadian science and technology
are mostly outsiders, at the periphery or beyond of their
discipline or profession. It was, in fact, a conference
of outsiders.

This fringe character of the study of the history of
Canadian science and technology contrasts vividly with the
encouragement and even enthusiasm with which the proposal
for the conference was greeted. Nowadays there are far
more complaints about the difficulties of getting funds for
conferences, research, or publication than there are
expressions of satisfaction, yet this conference had no

trouble receiving funding from many sources, and even
after our needs had been met there were further offers for
help if publication or other expenses required it.
Obviously, the funding agencies for research and publication in Canada put a much higher priority on studies into
the history of Canadian science and technology than do
academic researchers. I do not think the funding agencies
are wrong.

In recent years Canadians have been bombarded with
pleas, like those from the Science Council and the Symons
Report, for more research directed towards understanding
how science and technology have affected our development.
The funds are available. This sort of research cannot
develop overnight. If anything, academic disciplines are
marked by their caution rather than their boldness; by
their reliance on standard accepted bodies of knowledge
and tests of validity rather than by their innovation and
adaptability. Accuracy and truthfulness, not imagination
or creativity, are and should be the basic academic tests.
The study of the history of science and technology produces
difficulties because it must by its nature bridge the two
cultures, be interdisciplinary, and link together at a
minimum an understanding of scientific and technological
development with the instruments of history and social
science. This is a challenge most professional Canadian
historians have yet to face, and until they do, the study
of the history of our science and technology will be done
by outsiders. It would be well also to remember that interdisciplinary work can go in two directions: scientists
can invade history as legitimately as historians invade
science.

The great strength of the conference was that it
brought together such an enormous range of individuals
interested in the history of Canadian science and technology. We now have met each other, and can look for support
in a Canada-wide "invisible university." The challenge now
is for us to produce worthwhile scholarship. There will
be two aspects to research: on the one hand study of the
development of ideas and knowledge; on the other hand study

of the impact of ideas and knowledge on Canadian society.
I suspect there is, even at this early stage, a danger that
one of these (probably the first) will become respectable
to the neglect of the other. What the conference proved,
and what should be remembered, is that most of the basic
groundwork still needs to be done. We need studies of
successful research, and of the failures, of industries
that have prospered, and of those that went broke; of
government departments that had an impact and of those
that have been forgotten. Only then will we know how we
got to where we are (or even where we are) and how we might
get to somewhere more desirable. We will all continue to
be outsiders until more of this basic work is done.

The conference was only a beginning. We have published the proceedings almost *in toto* to give an accurate
representation of the state of the field. The warts show.
Some papers are very good, some not up to that standard.
We have also included a bibliography, so that the published
proceedings as a whole will serve as a source book to
students and teachers interested in the field. I hope that
in five or ten years another conference will be held, at
which each paper will be based on many published studies,
and all present will look back to the beginnings, in the
lean period of 1978.

WHAT IS CANADIAN ABOUT THE HISTORY
OF CANADIAN SCIENCE AND TECHNOLOGY?

I

Y A-T-IL UNE HISTOIRE DES SCIENCES
ET DE LA TECHNOLOGIE CANADIENNES?

WHAT IS CANADIAN ABOUT SCIENCE
IN CANADIAN HISTORY?[1]
Trevor Levere

In sketching themes distinctive of the roots of
English Canadian science,[2] I shall for the most part re-
strict myself to nineteenth-century Upper Canada, now
grown into Ontario. The science of the north, and
especially the arctic, was as different from that of the
south as life on the northern frontiers was from life in
the relatively settled south. I shall focus on science
based south of Superior. I shall also preserve a distinction between science and technology that, given the
circumstances of Victorian Canada, is more an artificial
convenience than a reflection of the past. Our nineteenth-
century science was, at least in aspiration, more useful
than theoretically significant.

Upper Canadian scientists were amateurs who moulded
themselves rapidly into professionals, forming a national
style of science within the Empire, and depending heavily
on the scientific education and entrepreneurial ideology
of Scottish immigrants.[3] They confronted their new land
with science as a talisman whose power they took on faith.
Confrontation with the land and entrepreneurial ideology
were the two factors that contributed most to give science
its special place in Canadian society and culture, in ways
that I shall distinguish from the American experience.
The science that emerged was in turn significant in shaping
our society and culture. That, more than intellectual
brilliance, is its importance.

Visitors to the Cavendish laboratory in Cambridge in
its nineteenth-century heyday were struck by the brilliance
of the results obtained with slender material resources.
"Our people think, instead of relying on equipment," they
were told. The disparity between achievement and resources
was different in kind and greater in extent in nineteenth-
century Canada. Human resources were initially even more

14

slender than material ones: there were only a handful of
professional scientists in mid-century British North
America, at a time when the reign of the technocrat and
scientific civil servant had already begun in Britain and
Europe. A scientist in the England of 1850 could make a
sizeable income from commissions and consultations from
government, industry, and the law.[4] With the notable
exception of scientifically trained officers of the Royal
Artillery and the Royal Engineers, hardly anyone in this
land could then claim to make a living from science.
Even William Logan, creator of the Geological Survey, had
frequently in the early years to dip into his pocket to
continue his work.[5]

The bulk of early scientific studies was carried on
by amateurs of increasing knowledge and sophistication.
Provincial entomology, for example, was long a purely
honorary and amateur affair. As late as the early 1900s,
western entomology, faced with everything from cutworms to
grasshoppers in plague proportions, rested upon the
labours of four or five professionals, and a somewhat
larger number of industrious amateurs.[6] Even the first
Dominion Entomologist, James Fletcher, had to educate himself through a rapid transition from near-amateur to
polished professional.

In such circumstances, early Canadian achievements in
science are all the more impressive. The first task was
to map resources, identify and classify species, and explore the natural history of the land as a prelude to
settlement. It is well to recognize the energy, discipline,
and enthusiasm of the early scientific surveyors and obser-
vers -- such men as Logan and his successors in geology;
Richardson and the other naturalists and zoologists who
sailed with Franklin;[7] Riddell, Lefroy, and their brother
officers in the magnetic survey;[8] and Fletcher in entomology. Canadian investment in science during the nineteenth century and until the Great War of 1914-1918 was
minimal. What successes there were came largely from
individual enterprise and its efficient deployment.

Part of the explanation for these early successes lies in the continuity of scientific tradition provided by the Imperial connection. The American Revolution led early to pressure within the United States to develop indigenous science. In the late eighteenth century, for example, London was the international capital and centre for the manufacture of scientific instruments, including surveying instruments of all kinds. Surveying in the States, however, began with virtually home-made instruments that, in a relatively short space of time, gave way to increasingly precise instruments of domestic manufacture.[9] Canadian surveying relied from the outset on English instruments, so that it was able to begin accurately, and to continue so. Then again, British North America was able to draw initially on the Royal Engineers and the Royal Artillery, who were almost unique in providing a disciplined, organized, and scientifically educated body of men. Magnetic and meteorological observatories, for example, were first established and manned by the Artillery.[10] The contrast with the "thinkers and tinkers" of the States, and with the more civilian character of science in Britain, is striking. Another Imperial connection was through the British Association for the Advancement of Science, whose first meeting outside Britain was in Montreal in 1884, with less of the tension that accompanied Canadian meetings of the American Association for the Advancement of Science. Proposals made through the Royal Society by J. W. Dawson that Canada should initiate and spearhead an Imperial Geological Union came to nothing, but this attempt indicates how even nationalistic assertions within Canadian science could be based on an Imperial foundation.[11]

The most important scientific legacy from Britain was the ideology and educational practice of the Scots.[12] Within French-speaking Quebec in the nineteenth century, science was taught as a part of general culture, but scientific research and its application were little in evidence. The German influence on research and science in Canada was relatively insignificant until the end of the

Victorian era, and, when it came, was largely transmitted through the United States.[13] England came late on the international scene to scientific education, nor were the immigrant Empire Loyalists much in advance here. But Scottish immigrants came from a country that had a far more impressive educational record. Widely educated in science, the Scots had by the late eighteenth century developed a combination of entrepreneurial and scientific skills that underlay the work of the improving landlords in the industrial revolution. Their ideology saw science not only as a tool for the understanding of nature, but also as a tool for the enrichment of mankind through the development and exploitation of nature.[14]

This ideology found a ready application in Canada. Immigrants were faced with the brute physical facts of the land, its vastness, its bounty and harshness, its extremes of climate and terrain. As we have seen, a first necessity was a portrait of the land with an inventory of resources -- hence the prominence by the mid-nineteenth century of the geological survey. Knowledge of the soil and of minerals led to at least a theoretical knowledge of where immigrants should be encouraged to settle, and of the requisite arteries of transportation that would make settlement viable. Sanguine prospectors could use geological maps, together with manuals like Henry White's *Gold: How and Where to Find It!*,[15] railroad engineers and scientists in the Palisser expedition were instructed to exchange scientific information, experimental farms grew up along the western railroads, and farmers were urged to make use of soil analyses in developing their land. Fertile soils around Edmonton, coupled with a latitude that would have been temperate in Europe, appear to have been promoted with the questionable authority of the Geological Survey to tempt immigrants to Edmonton's balmy farmlands.[16]

Geology was a fundamental science in farming, mining, and the development of transportation. It was a science that aroused great enthusiasm in Canada -- it was economically important, and enabled Logan to assure a review committee that science led to economics, and

economics led to science.[17] It was a science relatively
new in the 1840s,[18] and one to which the Canadian survey
could accordingly make significant contributions. It was
capable of conferring benefits on so many distinct activities that it became *the* Canadian science of the nineteenth
century, a vital part of a liberal and practical education
as it came to be conceived.

The need to come to terms with the land, to dominate
and exploit it instead of being overwhelmed by it, and the
evident success of geology in helping immigrants win that
confrontation, joined with the ideology of the Scottish
Enlightenment in creating a faith in science in nineteenth-
century Canada far stronger than that in Britain and
Europe. Egerton Ryerson in 1842 defined physical science
as "the knowledge of nature applied to practical and useful
purposes," and stressed its centrality to a liberal education.[19] The Toronto Library Association in 1854 was heated
to a discourse on "The Importance of Scientific Studies to
Practical Men," and in the following year Dawson, McGill's
new president, gave his views "On the Course of Collegiate
Education, Adapted to the Circumstances of British North
America." He too stressed science in practical education;
one of his proposals was for a course of commercial education, including "English Literature, History and Physical
Geography, Mathematics, Chemistry, Natural Philosophy,
Natural History, Modern Languages, Commercial Law; and
. . . Political Economy."[20]

Science, in short, as an article of belief, was held
to be culturally, practically, and economically valuable.
We have already encountered Logan's strategy -- geology
should be supported as an economic science. When the
British Association for the Advancement of Science visited
Montreal, almost every locally produced contribution
stressed utility -- entomology was practical and economic,
so were chemistry, meteorology, and a list of other
sciences.[21] Such labels reflected ideology rather than
true utility. It was all very well to urge a farmer to
analyze his soil -- but settlers were often more concerned
with removing boulders and stumps from their fields.

Worse yet, mid-nineteenth-century agricultural chemistry
was simplistic to the point of frequent error. There was
in Quebec a popular French translation of Humphry Davy's
Elements of Agricultural Chemistry--but by mid-century
this work was outdated, and some who tried its advice
found that they were killing their crops, not fertilizing
them. Davy's work promised readers that science was
useful; in the event, that promise was more significant
than performance.[22]

Since practicality seemed at once motive and reward for
the pursuit of science, the most successful approaches were
those stressing practice over theory. There were no objections to the pursuit of experimental science, so long
as the experiments were directed towards practical rather
than theoretical ends -- witness, for example, this advice
from the *Canadian Agriculturist* of 1850 to the University
of Toronto: ". . . whatever is attempted should bear upon
the face of it, the stamp of *practical utility*. The . . .
merely pointing out the application of some of the laws
and doctrines of chemistry, geology, animal and vegetable
physiology, etc., to the pursuits of the farmer, however
interesting and suggestive as many of these undoubtedly
are, would be quite a different thing from the practical
teaching of agriculture as *art*. The principle on which a
Professorship of Agriculture should be founded in the
present day, according to our notion, is that of *Practice
with Science*."[23] This approach was pursued with a fair
degree of consistency at every level of the educational
system, and was supported by government, the major consumer
of science in this country throughout the nineteenth century. Canadian science was organized and utilitarian;
the questions it tackled were distinguished by the practical solutions demanded.

With such an emphasis upon practice, it is not surprising that, outside geology, nineteenth-century Canadian
scientists tended to be cautious, indeed conservative.
The reaction to Darwin's theory of evolution is in this
respect typical. In Europe and the United States the
theory was widely debated, arousing instant interest

and controversy, with a good number of leading scientists
ranging themselves on either side. In Canada, the initial
reaction was negligible. When the theory came to be discussed, it was usually cautiously, critically, and nega-
tively.[24] Sectarianism within the churches reinforced
this tendency; it seems probable that scientific appointments in the universities were determined as much on
religious grounds as on scientific ones. Otherwise Thomas
Henry Huxley and John Tyndall, for example, might both have
been appointed at Toronto.[25]

Science, however, in spite of sectarianism, was more
a unifying than a divisive force in Canadian history. When
Logan organized Canadian geological exhibits of economic
minerals for the Great Exhibition of 1851 in London and
the Paris Exhibition of 1855, the recognition he received
abroad was seen as a national triumph at home. "Canada
Against the World!" was one banner headline in Upper
Canada.[26]

Canada, energetic, young, richly endowed, and guided
by an entrepreneurial scientific ideology, saw its future
wealth and place among the nations as fruits to be won by
science from nature. That science was deployed by an
optimistic bureaucracy, with an organization that complemented and underlined its utilitarian character. Geared
to the land, it recorded, tabulated, and mapped; Canadian
science was presented as a constructive tool of social
planning and social engineering, with a degree of seeming
promise and of government control distinctly exceeding the
contemporary norm among other nations.

NOTES

1. Cf. L. Lortie, "La trame scientifique de l'histoire du
   Canada," in *Pioneers of Canadian Science*, ed. G. F. G.
   Stanley (Toronto: University of Toronto Press, 1966),
   3-35.

2. Cf. R. Duchesne's paper in this volume, "Problèmes
   d'histoire des sciences au Canada français."

3. McGill and other English-speaking Montreal foci for scientific activity were, in the nineteenth century,
   closer to their fellows in Upper Canada than to the
   French Canadian scientific community.

4. M. Berman, *Social Change & Scientific Organization: The Royal Institution 1799-1844* (London: Heinemann, 1978).

5. B. Harrington, *Life of Sir William Logan* (Montreal, 1883). M. Zaslow, *Reading the Rocks: The Story of the Geological Survey of Canada* (Toronto: Macmillan, 1975).

6. Prof. P.W. Riegert is writing a history of entomology in western Canada.

7. J. Richardson, *Fauna-Boreali-Americana;or the Zoology of the Northern Parts of British North America*, 3 vols. (London, 1831).

8. J. H. Lefroy, *In Search of the Magnetic North:Soldier-Surveyor's Letters from the North-West, 1843-1844*, ed. G. F. G. Stanley (Toronto & New York, 1955). D. Thomson, *Men and Meridians*, vol. 1 (Ottawa, 1966).

9. E. G. R. Taylor, *The Mathematical Practitioners of Hanoverian England, 1714-1840* (Cambridge, 1966). S. Bedini, *Thinkers and Tinkers* (New York: Scribners, 1975).

10. T. H. Levere, "Edward Sabine," *DCB*, forthcoming.

11. J. W. Dawson, *Proceedings and Transactions of the Royal Society of Canada*, 1887, vi-vii, xiii.

12. A. Donovan, *Philosophical Chemistry in the Scottish Enlightenment* (Edinburgh, 1975).

13. A useful source is M. W. Rossiter, *The Emergence of Agricultural Science:Justus Liebig and the Americans, 1840-1880* (New Haven: Yale University Press, 1975).

14. T. H. Levere and R. A. Jarrell, *A Curious Field-book: Science and Society in Canadian History* (Toronto: Oxford University Press, 1974).

15. H. White, *Gold:How and Where to Find It!* (Toronto, 1867).

16. H. Y. Hind, *The Corruption of the Geological Survey in North-West Territory Matters*, 1883. See n. 14 above.

17. *Report of the Select Committee on the Geological Survey of Canada* (Quebec, 1855).

18. R. Porter, *The Making of Geology. Earth Science in Britain, 1660-1815* (Cambridge: University Press, 1977). Geology was of major importance throughout British North America; Gesner's work may be cited as a major contribution to the science in the Maritimes.

19. Levere & Jarrell, *Field-book*, 201.

20. *Ibid.*, 202, 206, 208.

21. *Report of the British Association for the* Advancement *of Science* (Montreal, 1884).

22. N. Aubin, *La chimie agricole mise à la portée de tout le monde* (Québec, 1847), was perhaps a more suitable, more modest, and safer guide -- but its ideology was very much in line with Davy's.

23. *Canadian Agriculturist* 1850, 265-6.

24. An exception is the work of J. W. Dawson, which was vigorously critical of Darwin's theory. J. W. Dawson, *Modern Ideas of Evolution*, ed. W. Shea & J. F. Cornell (New York: Science History, 1977).

25. A. B. Macallum, "Huxley and Tyndall," *University of Toronto Monthly*, 1901-02, 68-76.

26. *The Canadian Statesman*, Bowmanville, 6 September 1855.

PROBLEMES D'HISTOIRE DES SCIENCES

AU CANADA FRANCAIS

Raymond Duchesne

En 1885, une nouvelle épidémie de petite vérole ravage Montréal. Dans cette seule année, la maladie emporte au-delà de 3,000 personnes, ce qui, pour une ville d'au plus 150,000 habitants, représente une terrible saignée.[1] Mais il n'y a pas que l'ampleur du désastre qui risque de surprendre l'historien. Si l'on examine les statistiques de l'épidémie, telles qu'elles sont présentées dans le *Rapport annuel du Bureau d'hygiène de la Ville de Montréal*,[2] un fait étonnant se dégage; la maladie ne semble pas avoir frappé indistinctement la population montréalaise. En effet, le taux de mortalité, donné par 1000 habitants, augmente beaucoup plus dans les quartiers francophones de Saint-Louis, Saint-Jacques, et Sainte-Marie, passant de 29 à 61, en 1885, que dans les quartiers anglophones de Sainte-Anne ou de Saint-Laurent, où il ne passe que de 20 à 25. Données cette fois selon les groupes ethniques qui forment la population de Montréal, les statistiques établissent que le nombre des décès par 1000 habitants au cours de la terrible année 1885 est deux fois plus élevé chez les Canadiens français que chez les Irlandais et trois fois plus élevé que chez les montréalais d'origine anglaise ou écossaise. Autrement dit, la petite vérole a relativement épargné la population anglophone de la ville et la majorité des 3000 victimes de l'épidémie de 1885 ont été des Canadiens français.

Pour expliquer cet étrange développement de l'épidémie, nos ancêtres du 19e siècle, et après eux les historiens de la médecine comme, par exemple, Maude Abbott,[3] n'ont pas manqué de faire remarquer que la pratique de la vaccination était déjà généralisée au sein de la population anglophone alors qu'un projet de loi visant à la rendre obligatoire était constamment reporté à cause de l'opposition de la pop-

ulation canadienne-française. La chronique de l'époque
rapporte également qu'alors que les médecins anglais de
Montréal, avec à leur tête les professeurs de la Faculté
de Médecine de l'Université McGill, se faisaient les ar-
dents défenseurs de l'immunisation de la population par la
vaccination, leurs confrères canadiens-français présent-
aient aux charlatans et aux retardataires un front moins
uni. En fait, il fallut cette terrible épidémie de 1885
pour vaincre, non seulement dans la population, mais aussi
dans le corps médical francophone de Montréal, les dernières oppositions à la vaccination.

Mais cette différence des points de vue sur les dangers et les mérites de la vaccination n'est pas la seule
chose que distingue, dans la deuxième moitié du 19e siècle,
les médecins anglais du Québec et leurs confrères francophones. Ainsi, par exemple, la pratique des dissections et
l'étude de l'anatomie pathologique ne rencontrent pas du
tout la même faveur. Alors qu'à l'Université McGill,
William Osler, qui avait été en Europe l'élève des Virchow
et des Rokitansky, pratique une centaine de dissections
chaque année, où il puise la matière de son cours d'anatomie pathologique et des forts volumes annuels du *Montreal
General Hospital Pathological Report* et à partir desquelles
il rassemble une collection unique de pièces pathologiques,
l'Ecole de Médecine de Montréal et la Faculté de Médecine
de l'Université Laval doivent encore réclamer des autori-
tés civiles et religieuses qu'on leur accorde annuellement
un nombre suffisant de corps et se plaignent que leurs
étudiants font leur possible pour éviter de pratiquer des
examens post-mortems, souvent avec succès.[4]

Autre différence entre les médecins anglais et les
médecins français, plus suprenante et plus importante celle-
là; les méthodes aseptiques et antiseptiques développées par
Lister en Ecosse, dès 1865, et par les disciples de Pasteur
peu après, sont introduites beaucoup plus rapidement au Canada anglais que chez les médecins francophones. Alors qu'on
reconnaît à Sir Thomas Roddick, un disciple canadien de
Lister et un professeur de l'Université McGill, le mérite
d'avoir introduit, vers 1877, ces réformes majeures de la

chirurgie et de la médecine au Montreal General Hospital,[5]
ce n'est pas avant 1890 que celles-ci s'imposent défini-
tivement dans les salles de l'Hôtel-Dieu de Québec et de
l'Hôtel-Dieu de Montréal.[6]

Au total, les médecins canadiens-français de la seconde moitié du 19e siècle, à la différence de leurs confrères anglophones, ne sont pas convaincus de l'importance
de l'immunisation par vaccination, n'ont généralement, au
sortir de la faculté, qu'une connaissance rudimentaire et
toute livresque de l'anatomie et de la physiologie, ainsi
que de leurs évolutions pathologiques, et, enfin, n'observ-
ent souvent, au contact des malades et en s'approchant du
bloc opératoire, que les règles élémentaires de l'hygiène
et de la propreté.

Mais que les uns et les autres puissent également se
dire médecins, au même moment et au même endroit de l'histoire, en dépit de telles divergentes de vue et de telles
différences dans la pratique thérapeutique ne doit pas nous
surprendre outre mesure: après tout, les uns et les autres
ne servent pas la même population, ne pratiquent pas dans
les mêmes hôpitaux, n'appartiennent pas aux mêmes associations professionnelles et n'apprennent pas leur métier dans
les mêmes écoles. A côté des institutions anglophones du
Québec, le Montreal General Hospital, la Faculté de Médecine
de l'Université McGill et la Canadian Medical Association,
sur lesquelles s'appuient les médecins anglais, on trouve
un réseau complet d'institutions françaises et catholiques,
les Hôtel-Dieu de Québec, Montréal et Trois-Rivières,
fondés et administrés par des communautés religieuses,
l'Ecole de Médecine et de Chirurgie de Montréal, la Faculté
de Médecine de l'Université Laval et, à partir de 1876, sa
Succursale de Montréal, ainsi que le Collège des Médecins
et Chirurgiens de la Province de Québec, au service de la
population francophone du pays.

Au bout du compte, il nous faut reconnaître que, dans
la seconde moitié du 19e siècle, il existe au Canada deux
types distincts de pratique médicale, deux médecines, qui
diffèrent à la fois au plan du savoir appliqué et enseigné
et au plan des institutions où elles se pratiquent. Qu'elles

différent également au plan de leurs résultats, que le
choléra asiatique ou la petite vérole fauchent davantage
de Canadiens français chaque année, ou que même après 1880,
les fièvres d'hôpital, c'est-à-dire les infections post-
opératoires, continuent d'emporter la majorité des malades
dans les hôpitaux francophones, quelle que soit la dex-
térité du chirurgien, cela donne une juste idée de la dis-
tance qui les sépare et du peu cas que l'on fait, dans
chaque camp, de la science ou de l'ignorance du voisin, de
ses succès ou de ses échecs. Aussi suprenant que cela
puisse paraître, nos ancêtres du siècle dernier ne semblent
pas s'être rendus compte qu'ils se trouvait devant deux
médecines différentes et que l'une guérissait un peu mieux
que l'autre.

Le petit exposé que je viens de vous faire sur la
médecine au Québec à la fin du 19e siècle, en plus de faire
ressortir les mérites de l'histoire comparative, nous indique également quelque chose du statut général de l'histoire de la médecine au Canada comme objet d'étude. Le fait
que l'on puisse distinguer si clairement une médecine française et une médecine anglaise à un moment particulier de
notre histoire nationale, nous amène à reconnaître à chac-
une un statut propre dans l'histoire de la médecine au
Canada. Autrement dit, cette histoire de la médecine canadienne sera double; elle se fera à partir d'une histoire
de la médecine au Canada français et d'une histoire de la
médecine au Canada anglais, selon leur évolution respective
et en fonction des rapports qu'elles entretiennent.

Cette manière d'envisager le partage de l'histoire
médicale canadienne, qui, après tout n'est pas nouvelle, ne
devrait ni surprendre, ni troubler celui que sait à quel
point la pratique de l'art médical est déterminée par les
conditions sociales et culturelles de la société où elle
s'exerce. Le partage de la médecine canadienne, que nous
avons examiné au 19e siècle, mais qui est déjà sensible au
18e siècle, suit les frontières ethniques qui déchirent la
société canadienne depuis 1760, et qui sont de même frontières linguistiques, géographiques, religieuses et écono-

miques, et en un mot, traduit simplement la dualité de la
nation canadienne. Le chemin que suit le développement de
la médecine au Canada français est tracé par l'évolution
sociale et politique de la nation canadienne-française, par
des événements historiques qui marquent cette évolution;
et le travail de l'historien ne consiste pas seulement à
déterminer dans quelle mesure ce chemin diverge de ceux
qu'empruntent au même moment la médecine du Canada anglais
ou la médecine en Allemagne, mais aussi à expliquer pourquoi et comment, en dépit de cette divergence et de son
retard, la médecine du Canada français s'impose à son
milieu. On peut trouver aberrant ou scandaleux qu'encore
en 1885, dans une ville où William Osler a enseigné pendant
dix ans, une large partie de la population ne soit pas
vaccinée, ou que les succès de Roddick au Montreal General
Hospital ne changent rien au sort des femmes canadiennes-
françaises que les fièvres puerpérales continuent d'emporter jusqu'après 1890; si l'on ne rapporte pas la médecine à son contexte social, on ne dépassera pas de sitôt le
stade de l'étonnement et de l'indignation.

Mais ce statut propre que l'on accordera peut-être
sans trop de difficultés à l'histoire de la médecine au
Canada français, l'étendra-t-on pareillemnnt à l'histoire
des sciences et à l'histoire de la technologie? Est-on prêt
à admettre que, non seulement la science et la technologie
ont connu des développements différents au Canada français
et au Canada anglais, mais aussi que ces différences sont
si profondes, si importantes qu'elles justifient l'éclate-
ment de l'histoire canadienne des sciences et de la technologie en deux rubriques plus ou moins complémentaires,
l'une française et l'autre anglaise? Evidemment, pour qu'on
se résoude à pratiquer une telle opération, il faudrait que
l'on fasse apparaître, au plan des discours scientifiques
et au plan de l'histoire des institutions, certaines de ces
"importantes et profondes différences." Il faudrait, par
exemple, montrer qu'au Canada français et au Canada anglais,
le développement scientifique et technologique ne répond
pas aux mêmes besoins économiques ou sociaux, que les doc-

trines scientifiques n'ont pas les mêmes résonances idéologiques et qu'en conséquence, elles ne rencontrent pas
partout la même opposition ou la même faveur. Il faudrait
montrer aussi pourquoi ce ne sont pas les mêmes groupes ou
classes sociales qui, dans un contexte ou dans l'autre, se
chargent d'appuyer financièrement et politiquement le développement des sciences ou des techniques. Bref, il faudrait montrer que, pour la science et la technologie comme
pour la médecine, la réalité sociale du Canada, bi-ethnique
et bi-culturelle, détermine des développements conceptuels
et institutionnels différents et dont l'histoire doit res-
pecter l'originalité.

Rassurez-vous, je n'ai pas l'intention de démontrer
tout cela maintenant: nous n'en sommes encore, au plan de
la définition de notre champ d'étude, histoire des sciences
au Canada ou histoire des sciences au Québec, qu'au stade
des hypothèses, des préjugés et du parti-pris. Cela admis,
je vais me contenter de citer deux très courts exemples
afin d'illustrer la nécessité d'une sécession de l'histoire
des sciences au Québec. Ils sont tous deux tirés du 19e
siècle et font appel à des éléments relativement connus de
notre passé scientifique.

Ceux qui se sont donnés la peine de faire des com-
paraisons[7] savent déjà qu'au 19e siècle, la bourgeoisie du
Canada anglais et celle du Canada français ne perçoivent
pas les sciences d'une manière identique et qu'elles ne
ressentent pas également l'importance d'aider à leur professionnalisation et à leur institutionnalisation. Alors
que la bourgeoisie d'affaires anglophone se montre particulièrement intéressée par l'aspect utilitaire de la science et
par ses éventuels développements technologiques, la bourgeoisie canadienne-française, où dominent les membres du
clergé et des professions libérales, envisage d'abord les
études scientifiques comme un complément heureux à la formation culturelle du jeune homme de la "bonne société."
Cette différence de point de vue est cruciale dans le processus d'institutionnalisation et de professionnalisation
de la science; l'enseignement des sciences que l'on dispense
à l'Université McGill et à l'Université de Toronto trouve

son application et du même coup sa légitimation dans les entreprises du Geological Survey, des Dominion Experimental Farms et du Biological Board. Au Canada français, l'institutionnalisation des disciples scientifiques ne dépasse guère, au 19e siècle, les quelques cours que l'on dispense à l'Université Laval et dans certains collèges classiques. En outre, l'histoire naturelle que pratique l'abbé Léon Provancher et les quelques amateurs qu'il a rassemblés autour de son journal, le *Naturaliste canadien*, se démarque assez radicalement, au plan des hypothèses et de la doctrine, du travail des naturalistes canadiens-anglais;[8] en effet, les premiers semblent ne s'intéresser qu'au jeu de l'identification taxonomique des espèces botaniques et entomologiques présentes dans la vallée du Saint-Laurent, et ignorent superbement la discussion des principes de la classification, la physiologie et l'anatomie, les hypothèses transformistes qui se multiplient après 1859 et le débat qu'elles suscitent, l'application de la botanique et de l'entomologie au développement de l'agriculture canadienne, etc.... Lorsqu'on la compare à l'histoire naturelle qui se pratique alors au Canada anglais et aux Etats-Unis, la science de l'abbé Provancher apparaît sinon comme une aberration de la pensée scientifique, du moins comme une doctrine qui retarde énormément sur les idées du jour et qui souffre beaucoup de ne pas s'être soumise davantage à la critique des pairs. Mais si on la reporte à son contexte social, où elle n'est qu'un vernis culturel que l'on a ni l'occasion d'appliquer dans la vie quotidienne, ni l'occasion de confronter à la science de correspondants ou de visiteurs étrangers, l'histoire naturelle de l'abbé Provancher, fixiste et nationaliste, s'explique bien sa longévité, dans un univers scientifique rendu fluide par les progrès de la biologie évolutionniste, n'a rien qui doive surprendre.[9]

Pour donner un dernier exemple de l'importance des facteurs sociaux et culturels dans le développement de la science au Canada et, par conséquent, de la nécessité de distinguer l'histoire des sciences au Québec de l'histoire des sciences au Canada anglais, rappelons la création, en

1882, de notre grande académie scientifique nationale, la
Société royale du Canada, à laquelle participent quelques
savants du Canada français. On a avancé récemment la thèse
selon laquelle la Société royale aurait été fondée, non
pas tant pour servir la professionnalisation de la science
au Canada, que pour exprimer, sur la scène scientifique
internationale et sur la scène politique locale, le nation-
alisme canadien, exacerbé à la fois par le dépendance
coloniale et par l'impérialisme économique et culturel
américain.[10] Les arguments présentés à l'appui de cette
thèse sont pour la plupart tirés de la situation politique
et du climat idéologique qui prévalent au Canada vers 1880,
de même que des témoignages rendus par les scientifiques
canadiens-anglais de l'époque; ils ne nous apprennent donc
pas grand chose des motifs qui ont incité de savants canadiens-français à se joindre à leurs collègues anglophones.
Mais, même en supposant que des personnages comme Charles
Baillargé, Monseigneur Laflamme et l'abbé Provancher se
soient associés aux premiers travaux de la Société parce
qu'ils partagent avec les Dawson et les George Lawson la
même fierté d'être canadiens, ce qui n'est pas prouvé,[11]
il n'est pas certain qu'ils aient été en cela très repré-
sentatifs du sentiment de la population québécoise, ni même
des besoins propres à la petite communauté scientifique du
Canada français. Chose certaine, l'exemple de la Société
royale du Canada nous donne à penser que, même lorsqu'ils
s'associent dans des entreprises communes, les scientifiques
canadiens-français et canadiens-anglais ne le font pas
nécessairement pour les mêmes raisons et, par conséquent,
traduisent pas la même réalité institutionnelle ou sociale.

Je ne crois pas devoir répéter, en guise de conclusion,
les arguments qui militent en faveur d'une déclaration
d'indépendance de l'histoire de la médecine, des sciences
et de la technologie au Canada français. Les difficultés
sont telles qu'il faudra bien, un jour ou l'autre, abandon-
ner l'ambition d'écrire cette histoire unique de la science
canadienne qui ne reconnaît pas la spécificité de l'expérience québécoise et dont l'interrogation centrale, au sujet
du Canada français, est d'expliquer pourquoi on y est tou-

jours en retard d'une idée, d'une découverte ou d'un prix
Nobel sur le Canada anglais et comment il se fait que les
Canadiens français n'aient pas fourni leur juste part à
l'effort scientifique et technologique national.

On reconnaîtra plus sûrement cette histoire tradition-
nelle si l'on ajoute qu'on y fait invariablement le procès
de l'antimodernisme du clergé catholique ou du manque
d'initiative économique et technologique du Canadien français, quand ce n'est pas celui d'un manque d'aptitude con-
génital pour les hautes abstractions de la science.[12] Outre
qu'elle n'explique pas grand chose, cette histoire "unifi-
ante" passe toujours si près d'écorcher les susceptibilités
ethniques qu'elle risque d'avoir des effets désastreux
pour l'unité nationale....

Les historiens de la science et la technologie au Canada ont mieux à faire que de reprendre ces clichés; les
historiens de la science et de la technologie au Québec
n'ont pas à se porter à la défense d'un honneur national
outragé; dans la mesure où est explicité le rapport de
l'histoire québécoise à l'expérience canadienne, des tâches
nouvelles et plus importantes attendent les uns et les
autres.

NOTES

1. Le nombre des décès imputables à la variole nous est
   donné par Maude E. Abbott, *History of Medicine in the
   Province of Quebec*, Montréal, McGill University, 1931,
   p. 62. Pour connaître la population de la ville à
   cette époque, on pourra consulter Jean-Claude Marsan,
   Montréal en évolution, Montréal, Fides, 1974, p. 308.

2. Ces statistiques sont citées, pour la période s'étendant
   de 1876 à 1896, par Jacques Bernier, "La condition des
   travailleurs, 1851-1896," dans *Les travailleurs
   québécois 1851-1896*, Montréal, Presses de l'Université
   du Québec, 1975, pp. 31-60.

3. *Ibid.*

4. Ainsi, Charles-Marie Boissonnault, dans son *Histoire de
   la Faculté de Médecine de Laval*, Québec, Presses de
   l'Université Laval, 1953, raconte que, pendant quelques années, les étudiants de Laval venaient faire
   leur dernière année de médecine à la Succursale de
   Montréal où on ne les obligeait pas à pratiquer de
   dissection.

5. Sir Thomas Roddick fit plusieurs voyages à Edimburgh
   pour observer le travail de Lister, mais ce n'est
   qu'en 1877 qu'il commence à utiliser le "carbolic
   spray" en chirurgie et ce n'est que deux ans plus
   tard qu'il publie un rapport sur les premières
   opérations qu'il a faites en observant les nouvelles
   méthodes prophylactiques. Cf. H. E. MacDermot, *One
   Hundred Years of Medicine in Canada, 1867-1967*,
   Toronto, McClelland & Stewart, 1967, p. 29.

6. Voir à ce sujet L. D. Migneault, "Histoire de l'Ecole
   de Médecine et de Chirurgie de Montréal," *Union
   médicale* 55 (9)(1926), pp. 597-674, et C. M. Boissonnault, *Faculté de Médecine*.

7. Je pense ici tout particulièrement à l'article de
   Richard A. Jarrell, "The Rise and Decline of Science
   at Quebec 1824-1844," *Histoire sociale* (1977), pp.
   77-91.

8. Sur la science de l'abbé Provancher, on pourra lire
   Jacques Rousseau et Bernard Boivin, "La contribution
   à la science de la *Flore canadienne* de Provancher,"
   Naturaliste canadien 6 (95) (1968), pp. 1499-1530, de
   même que G. P. Holland, "L'abbé Léon Provancher, 1820-
   1892," dans G. F. G. Stanley (éd.), *Pioneers of
   Canadian Science*, Toronto, University of Toronto Press,
   1966, pp. 44-53. On y trouvera deux évaluations
   critiques de l'oeuvre de Provancher en botanique et
   en entomologie, les deux disciplines qu'il a le mieux
   maîtrisées.

9. En fait, les fondements idéologiques de la pratique de
   l'abbé Provancher se maintiennent à peu près jusqu'en
   1920, lorsque la vigoureuse poussée du Frère Marie-
   Victorin et surtout une demande sociale accrue pour
   le savoir scientifique et technique viennent révolu-
   tionner l'enseignement et la pratique de la science
   au Canada français.

10. Peter J. Bowler, "The Early Development of Scientific
    Societies in Canada," dans Alexandra Oleson et S. C.
    Brown (éds.), *The Pursuit of Knowledge in the Early
    American Republic*, Baltimore, Johns Hopkins University
    Press, 1976, pp. 326-339.

11. On sait en tout cas que l'abbé Provancher était d'un
    nationalisme français farouche, comme en témoignent
    les articles politiques qu'il publiés dans divers
    journaux de son temps et comme le rapporte son bio-
    graphe; Mgr V. A. Huard, *La Vie et l'Oeuvre de l'Abbé
    Provancher*, Québec, Garneau, 1926.

12. On peut déplorer ici que la seule vue d'ensemble de
    l'histoire technologique canadienne appartienne juste-
    ment à cette tradition historiographique; J. J. Brown,
    Ideas in Exile. A History of Canadian Invention, Tor-
    ronto, McClelland & Stewart, 1967.

OPPORTUNITIES AND POTENTIAL IN
ARCHIVAL RESOURCES
II
LES ARCHIVES:
RESSOURCES ET POTENTIEL

INTRODUCTION

The relationship between archivists and their
clients--be they historical geographers, historians, or
journalists--is often a most curious one. Research is the
foundation of historical writing and yet few historians
have either knowledge of, or interest in, how others decide
what shall ultimately be saved for their use. Too often
influential researchers are publicly uninterested in the
processes whereby others decide what is or is not to be
grist for the historical mill. Many archivists, on the
other hand, fail to give more than lip service to the
idea of trying to adjust policies and priorities to
changing historical trends and societal concerns. Very
little Canadian archival activity has been carried out in
the name of advancing the cause of research in the history
of Canadian science, technology, or medicine. This blind
spot notwithstanding, there are some very fine archival
collections which deserve further research. Two such
collections are introduced here by Professor Ray and
Mr. Gillis.

All three papers have indicated that at present there
is no lack of significant questions to be put to existing
collections. There are, as Prof. Bernier has pointed out,
research difficulties but at the present there is more
material than workers. But if the research interest and
activity grows at the rate so confidently predicted by
many delegates at the Kingston conference, will there not
be a need for greater coordination between researchers and
the archivists who decide what is and is not to be the raw
material of Canadian history? Is it time to move out of
good luck and into good management? The fact that few were
aware of the potential in the collections discussed by Ray
and Gillis illustrates that not all of the Canadian solitudes are linguistic.

INTRODUCTION

Le lien entre archivistes et usagers, que ces derniers soient des spécialistes de la géographie historique, des historiens ou encore des journalistes, est souvent très curieux. L'énonciation historique repose sur la recherche; pourtant il est peu d'historiens qui savent ou même s'intéressent à la façon dont les archivistes décident de ce qui doit être préservé pour mener à bien leurs travaux. Trop souvent, des chercheurs de premier plan se déclarent peu intéressés par les procédures en vigeur qui régissent la constitution des stocks de matières premières nécessaires à l'entreprise historique. De l'autre côté de la barrière, de nombreux archivistes sont plutôt réticents à l'idée d'ajuster leurs manières d'agir et leur priorités en fonction de l'évolution des tendances historiques et des centres de la société. En ce qui concerne l'histoire des sciences, des techniques et de la médecine canadiennes, les archives du pays n'ont pas accompli grand-chose. En dépit de cet aveuglement, il existe pourtant quelques collections d'archives excellentes et qui méritent d'être exploitées à l'avenir. M. le professeur Ray et M. Gillis nous présentent ici deux collections de ce genre.

Les trois communications de cet atelier montrent que, pour le moment, les questions significatives à poser sur la base des archives disponibles ne manquent pas. Comme le souligne M. le professeur Bernier, des difficultés matérielles entravent les recherches, mais le nombre des chercheurs demeure bien limité quand on le compare à la taille des archives. En tout cas, si l'intérêt pour ce type de recherches et de travaux devaient se multiplier au rhythme qu'on prédit avec assurance de nombreux participants au congrès de Kingston, ne doit-on pas envisager d'harmoniser les exigences des chercheurs avec ce que les archivistes sont en mesure d'offrir dans la mesure où ce sont ces derniers qui, encore une fois, décident ce qui sera retenu et ce qui sera rejeté dans la constitution de la matière première de l'histoire canadienne? Le moment est-il enfin arrivé, où des mesures de saine gestion se substitueront au hasard? Peu de participants étaient au courant de la richesse des archives décrites par Ray et Gillis et ceci suffit à illustrer que les solitudes canadiennes ne sont pas seulement d'ordre linguistique.

L'HISTOIRE DE LA MEDECINE QUEBECOISE AUX
XVIIIe ET XIXe SIECLES: PROBLEMES ET SOURCES
Jacques Bernier

L'histoire de la médecine connait actuellement au Canada un développement important. Des postes même ont été
créés dans certaines universités confirmant ainsi son statut. Le Québec reste cependant relativement marginal à ce
courant puisqu'aucun poste d'historien de la médecine québécoise n'a encore été créé dans une université francophone. Pourtant la province de Québec n'est-elle pas la
plus vieille province du Canada et n'en fut-elle pas numé-
riquement la plus importante jusqu'au milieu du 19e siècle?

Bien qu'elle soit jeune, l'historiographie québécoise
relative à la médecine n'est cependant pas inexistante, et
elle nous a valu de façon épisodique une littérature de val-
eur inégale. Il faut dire aussi que les connaissances
actuelles sur l'histoire de la médecine québécoise sont
grandement tributaires d'études plus générales portant sur
l'ensemble du Canada.

Fait particulier, une partie importante de cette littérature date des années 1920. Cette décennie nous a valu
entre autres des travaux documentaires qui restent encore
parmi nos meilleurs repères chronologiques. Je pense en
particulier aux travaux de Ahern,[1] Heagerty,[2] Abbott,[3]
Migneault,[4] et à l'étude plus ponctuelle de A. Vallée sur
M. Sarrazin et son époque.[5] Les années 1930 à 1960 sont
marquées, quant à elles, par un vide un peu étonnant. On
innove peu et plusieurs des publications de cette époque ne
font que reprendre sous d'autres formes les travaux des auteurs de la période précédente. Quelques exceptions cependant méritent d'être signalées: l'étude de Boissonneault
sur l'histoire de la faculté de médecine de Laval,[6] celle
de Gauthier sur l'histoire de la société médicale de Québec[7] et celles de MacDermot sur la Canadian Medical Association,[8] le Montreal General Hospital[9] et la vie médicale

au Canada de 1867 à 1967.[10]

Le courant historiographique le plus récent coincide avec les débuts de la Commission d'enquête sur la santé et le bien-être social dont les travaux furent publiés en 1970. Une prise de conscience s'effectue alors et c'est à ce moment que le Québec se donne ses premiers sociologues, administrateurs et théoriciens en politique de la santé. L'intérêt pour l'histoire de la médecine renaît et commence à donner lieu à des études qui dépassent le niveau événementiel. Notons, entre autres, les articles de Gilles Dussault sur la profession médicale de 1940 à 1960 et ceux d'André Paradis et de ses collaborateurs sur le psychiatrie de 1800 à 1880.[11] L'historiographie québécoise bénéficia aussi durant cette période des articles de Sylvio Leblond[12] sur le début du XIXe siècle[13] et d'études générales sur différents aspects de l'histoire de la médecine canadienne.[14] Enfin, une des réalisations les plus utiles des années 1970 aura été le *Dictionnaire biographique du Canada* qui permet de mettre en évidence de nombreuses personnalités, jusqu'ici fort mal connues, et qu'il aurait été très difficile de connaître autrement.

L'histoire de la médecine québécoise est donc, pour ainsi dire, encore à ses débuts. Heureusement de plus en plus de projets de recherche émergent dans divers instituts et départements de la province et d'ailleurs.

Pour être fructueuses et conduire à des résultats valables dans un proche avenir, ces recherches devraient tenir compte, à mon avis, de certaines exigences particulières. Il m'apparaît d'abord indispensable d'avoir une bonne connaissance des sources disponibles sans quoi nous risquons de passer à côté de plusieurs questions essentielles. Il m'apparaît important aussi d'élargir le débat et que ces recherches soient faites en concertation avec des spécialistes de formations différentes. Je pense en particulier aux démographes, aux historiens, aux médecins, aux sociologues et aux anthropologues.

Face à cette histoire à découvrir, quatre grands secteurs m'apparaissent s'imposer en premier lieu: l'histoire des maladies; l'histoire du savoir, de l'enseignement et des pratiques médicales; le développement de la médecine

comme profession; enfin l'histoire des institutions
hospitalières et de l'hygiène publique.

Pour diverses raisons, l'histoire des maladies qui
aurait pourtant constituer le premier champ d'analyse de
cette histoire de la médecine reste encore peu développé
chez-nous. Rien de comparable aux travaux d'Ackerknecht,
Rosenberg ou Chevalier n'a été fait ici pour le mal de la
Baie Saint-Paul, la variole, la tuberculose, etc.... Une
étude sur le choléra à Québec mérite cependant d'être
signalée, celle de G. Bilson, sur les attitudes face au
choléra à Québec en 1832.[15]

Il faut donc entreprendre d'écrire l'histoire des maladies et des causes de mortalité au Québec. De telles
études ne devraient cependant pas se contenter d'identifier
les maladies, d'en faire une chronologie et de déceler leur
impact démographique, économique et social.[16] Il est nécessaire aussi, à la suggestion du regretté G. Rosen,[17] qu'elles mettent en évidence les causes de leur apparition et
les facteurs qui ont rendu possible leur disparition.

L'étude de ces questions soulève cependant certaines
difficultés importantes. Un premier problème vient de ce
que la terminologie médicale, pour la période antérieure au
développement de la médecine moderne à la fin du 19e siècle,
diffère souvent de celle utilisée de nos jours. Le savoir
médical étant avant cette période peu homogène, il est parfois difficile de comparer des données entre une période et
une autre, entre une province et une autre, et même entre
une ville et une autre. Ainsi un des premiers travaux à
mettre en oeuvre serait de faire une étude sur l'évolution
de la terminologie médicale au Québec.

Une bonne étude de l'histoire des maladies m'apparaît
également illusoire sans une critique approfondie des séries
statistiques utilisées. Ces sources sont nombreuses et var-
iées. On en trouve entre autres dans les revues médicales,
dans les rapports des hôpitaux, dans les rapports des bur-
eaux sanitaires municipaux et provinciaux ainsi que dans
les publications gouvernementales. Le fait qu'elles furent
compilées par des auteurs différents et pour des motifs é-
galement différents rend parfois ces sources incompatibles
les unes avec les autres. D'où la nécessité de les utiliser

avec prudence.

En réponse à ces maladies, des hommes sont intervenus
faisant usage de pratiques. Un des premiers problèmes au-
quel se heurte l'historien qui s'intéresse aux pratiques
médicales avant la deuxième moitié du 19e siècle, tient
précisément à la diversité de ces pratiques et de leurs savoirs. Il n'y a pas à l'époque, et encore moins que nos
jours, de savoir unique et exclusif. Au contraire, plusieurs types de pratiques cohabitent souvent sur un même
espace. Ceci pour plusieurs raisons. D'abord parce que les
médecins sont peu nombreux dans les campagnes et que l'hab-
itude de recourir à leurs services n'est pas aussi répandue
que de nos jours. Avant de s'adresser au médecin, on pré-
fère plutôt avoir recours à l'auto-médication ou s'en re-
mettre aux "spécialistes" du village ou du quartier tels
les rebouteux, les herboristes, etc. De plus il faut se
souvenir qu'en plus d'exiger des honoraires élevés, le médecin ne jouit pas encore de la crédibilité qu'il aura à
la fin du 19e siècle. En ville, il est en plus contesté par
l'homéopathe qui jouit dans plusieurs milieux d'une recon-
naissance certaine. Cette diversité des pratiques suscite
évidemment plusieurs questions. Par exemple, qu'est-ce qui
distingue ces pratiques les unes des autres? et quelles
couches sociales touchent-elles? Il sera relativement facile de répondre à ces questions pour la médecine orthodoxe
et la médecine homéopathique parce que la littérature médicale de l'époque existe toujours et que nous pouvons re-
tracer, dans les archives notariées, certaines de leurs
clientèles. La réponse est plus difficile en ce qui concerne les pratiques populaires. Celles-ci n'ont ordinaire-
ment pas fait l'objet de traités, et de plus, la médecine
constituait souvent pour ces praticiens une activité secon-
daire.[18] Pourtant, des indices de ces pratiques peuvent
être connus grâce à certaines sources orales utilisées par
les spécialistes du folklore. Une équipe de chercheurs du
C.E.L.A.T. de l'Université Laval poursuit d'ailleurs actuellement des études prometteuses en ce sens.

La fin du 19e siècle marque aussi le développement
des professions médicales au Québec. Bien que celles-ci

fassent depuis quelques années l'objet de débats importants, peu a été écrit à leur sujet dans une perspective historique. Pourtant, il m'apparaît important de
connaître le contexte et les raisons qui ont rendu possible
le développement du professionalisme en médecine au Québec. Pour ce faire, on pourrait commencer par une étude
attentive de la législation et des débats auxquels elle a
donné lieu dans les journaux et dans les revues médicales.
Pour vérifier si le développement professionel a été un
moyen d'ascension sociale on pourrait, à l'aide des inven-
taires après décès, des Poll Books et de certains recense-
ments nominaux, faire l'analyse de l'évolution du niveuu
de vie des médecins comparativement à d'autres groupes sociaux. Une étude de l'évolution de leur participation à
la vie politique fédérale, provinciale et municipale serait
également possible grâce, entre autres, aux documents officiels.

Outre cette question de la professionnalisation,
l'histoire de la profession médicale québécoise m'apparaît
marquée par un autre phénomène important, et lui aussi fort
peu étudié: celui de l'impact de la conquête dans le secteur médical. La médecine québécoise francophone traverse
après 1760 une période difficile, si bien qu'au début du
19e siècle 73 pour cent des médecins du Bas-Canada sont
anglophones.[19] Les causes, à l'origine de cette situation,
sont nombreuses et complexes.

Un premier élément d'explication peut être attribu-
able au fait qu'après la conquête, les postes officiels et
plus rémunérateurs ont été attribués dans ce secteur,[20]
comme dans le reste de l'administration, à des Britanniques.
A cette éviction des postes-clefs s'ajoute aussi, d'après
les sondages que nous avons effectués, le fait que la plupart des médecins et chirurgiens britanniques qui vinrent
ici à la fin du 18e siècle avaient reçu une formation dans
les écoles et qu'ils jouissaient de ce fait d'une plus
grandes crédibilité au sein des élites. Aussi bénéficiaeint-
ils, dans les villes en particulier, d'une clientèle plus
vaste et plus fortunée. Par contre, peu de Canadiens pou-
vaient se permettre le luxe d'aller poursuivre de telles

études à l'étranger, et l'apprentissage auprès d'un médecin unique restait en pratique le seul moyen de formation.
Cette formation, pensons-nous, fut bientôt dépréciée de
sorte que les Canadiens virent leurs chances de succès de
plus en plus réduites au sein de cette profession et de ce
fait l'abandonnèrent. Mais ce sont là des hypothèses qui
demandent à être développées et vérifiées.

Outre l'élaboration de savoirs et la mise en place de
corps de praticiens spécialisés, les maladies ont aussi
provoqué le développement des dispensaires, des hôpitaux et
de ce que d'une manière plus générale on appelle le secteur
de l'hygiène publique. C'est-à-dire l'ensemble des mesures
publiques prises par la société pour prévenir et guérir les
maladies. Au Québec, ces services ont longtemps été laissés
au clergé en échange de concessions de terres et de divers
privilèges. Cette situation devait changer au 19e siècle
alors que l'on remarque une participation de plus en plus
grande de l'Etat. Une première question se pose ici: quelles
ont été les raisons qui ont amené l'intervention accrue de
l'Etat dans ce secteur tout au long du 19e siècle? La re-
crudescence des maladies et le caractère désuet de ces institutions ont certainement été des motifs importants à l'o-
rigine de ces changements. D'autres facteurs cependant me
semblent avoir joué eux aussi un rôle non négligeable. Le
fait, entre autres, que ce glissement survienne au même
moment que la médecine "se professionnalise" me semble sig-
nificatif. En effet, je pense qu'afin d'exercer un contrô-
le de plus en plus étroit sur la pratique médicale, la profession médicale a besoin de s'approprier la gestion des
services de santé de le pouvoir d'orienter les nouvelles politiques de ce secteur.

Outre ces questions plutôt vastes, il y a aussi des
sujets plus faciles qui, s'ils étaient étudiés, seraient
d'une grande utilité. Je pense en particulier à des mono-
graphes sur les hôpitaux. De telles études s'imposent pour
rendre possible des analyses plus générales sur le développement du système hospitalier au Québec et pour faire ressortir ses caractéristiques par rapport aux autres modèles
nord-américains. Ces études devraient s'intéresser non

seulement aux raisons de fondation, aux modes de financement et d'administration, au personnel embauché et aux
traitements dispensés, mais aussi à l'origine sociale des
malades, et à la fréquence et à la durée de leurs séjours.
Ce genre d'études serait relativement facile à faire, en
particulier pour la période postérieure au milieu du 19e
siècle, puisque plusieurs institutions ont gardé leurs
procès-verbaux, leurs registres d'entrée des malades et
leurs livres de comptabilité.

En somme, l'histoire sociale de la médecine est un
domaine passionnant et les sujets à l'étude ne manquent
pas. Certains dangers cependant nous guettent tous. D'abord,
comme l'a souligné Perkins, celui de vouloir trop embrasser
à la fois, "The social historian must avoid the attempt to
be everywhere at once."[21] L'autre est celui de l'isolement
et le danger de ne pas bénéficier suffisamment des apports
de nos collègues des sciences sociales et de la médecine.

Enfin, pour terminer, j'aimerais rappeler qu'il nous
manque un instrument de travail essentiel: un inventaire
des fonds d'archives relatifs à la médecine. Un tel ouvrage
serait nécessaire, non seulement pour accélérer les recherches, mais aussi pour nous permettre d'identifier de nouveaux problèmes dont nous ne soupçonnons peut-être même
pas l'existence aujourd'hui.

NOTES

1. M. J. et G. Ahern, *Notes pour servir à l'histoire de la
   médecine dans le Bas-Canada,* Québec, 1923.

2. J. J. Heagerty, *Four Centuries of Medical History in
   Canada,* Toronto, Macmillan, 1928, 2 vol.

3. M. Abbott, *History of Medicine in the Province of Quebec,*
   Montreal, McGill University, 1931.

4. L.-D. Mignault, "Histoire de l'Ecole de médecine et de
   chirurgie de Montréal," *L'Union médicale du Canada,*
   55 (1926).

5. A. Vallée, *Un biologiste canadien, Michel Sarrazin,
   1659-1735, sa vie, ses travaux et son temps,* Québec,
   Imprimeur du roi, 1927.

6. C. M. Boissonnault, *Histoire de la faculté de médecine
   de Laval,* Québec, Presses de l'Université Laval, 1953.

7. C. A. Gauthier, "Histoire de la société médicale de Québec," *Laval Médical*, 8 (1943).

8. H. E. MacDermot, *History of the Canadian Medical Association, 1857-1921*, Toronto, 1935.

9. H. E. MacDermot, A *History of the Montreal General Hospital*, Montreal, The Montreal General Hospital, 1950.

10. H. E. MacDermot, *One Hundred Years of Medicine in Canada, 1867-1967*, Toronto, McClelland and Stewart, 1967.

11. G. Dussault, "Les médecins du Québec 1940-1960," *Recherches sociographiques*, 16, no 1 (avril 1975); A. Paradis, "L'émergence de l'asile québécois au XIXe siècle," *Santé mentale au Québec*, 2, no 2 (novembre 1977); A. Paradis, *Essai pour une préhistoire de la psychiatrie au Canada, 1800-1885*, Montréal, Département de philosophie de l'U.Q.A.M. (Recherches et théories, no 15), 1977.

12. S. Leblond, "La médecine dans la province de Québec avant 1847," *Cahiers des Dix*, Montréal, 1970; S. Leblond, "La profession médicale sous l'Union, 1840-1867," *ibid*, Montréal, 1973.

13. Voir aussi M. Allard et autres, *L'Hôtel-Dieu de Montréal, 1642-1973*, Montréal, Hurtubise HMH, 1973.

14. D. W. Gullett, A *History of Dentistry in Canada*, Toronto, University of Toronto Press, 1971; G. H. Agnew, *Canadian Hospitals 1920 to 1970*, Toronto, University of Toronto Press, 1974; G. Wherrett, *The Miracle of the Empty Beds:A History of Tuberculoses in Canada*, 1977.

15. G. Bilson, "The First Epidemic of Asiatic Cholera in Lower Canada, 1832," *Medical History*, 21, no 4 (octobre 1977).

16. En particulier sur leur impact selon la classe sociale, le groupe ethnique, le sexe ou le métier.

17. G. Rosen, "Social Variables and Health in an Urban Environment: the Case of the Victorian City," *Clio Medica*, 8 (1973), p. 1.

18. Pour laquelle ils ne se faisaient du reste pas toujours payer car il aurait été inconvenant, selon la mentalité de l'époque, de recevoir de l'argent pour avoir exercé un don reçu de la Providence.

19. Voir à ce sujet: F. Ouellet, *Le Bas-Canada, 1791-1840*, Ottawa, Editions de l'Université d'Ottawa, 1976, p. 269.

20. Notamment chez les médecins militaires.

21. H. J. Perkins, "Social History," dans H. P. R. Finberg (éd.), *Approaches to History*, London, 1962, p. 59.

OPPORTUNITY AND CHALLENGE:
THE HUDSON'S BAY COMPANY ARCHIVES AND
CANADIAN SCIENCE AND TECHNOLOGY
Arthur J. Ray

Without doubt, the Hudson's Bay Company Archives contain the most important collection of documents dealing
with native history and Canadian economic development in
the pre-Confederation period. Between 1950 and 1967, these
records were microfilmed and the copies were deposited in
the Public Archives in Ottawa.[1] In 1975 the original manuscripts were transferred from London, England to Winnipeg,
Manitoba. Given the rich detail of these records, and
their increased accessibility to Americans and Canadians,
it is not surprising that the pace of native and fur trade
history was greatly stimulated during the past decade by
these two developments.

As the pace of research has quickened, its scope has
broadened considerably. For example, in the area of native
history, initially most attention was focussed on questions
of changing patterns of socio-political organization,
cultural ecology, tribal migrations, population change,
economic dependency, and material culture change. As
scholars dealt with these topics, they found that it was
necessary to go farther afield for useful insights. The
work of environmental scientists, medical researchers,
nutritional scientists, and historians of science and technology all became more relevant. At the same time, it has
become increasingly clear that the Hudson's Bay Company
Archives have great research potential for scholars in
these other disciplines. To date, however, few have recognized this potential and exploited it.

It is primarily in the area of environmental science
that important new work has begun. Two geographers,
D. W. Moodie and A. Catchpole, were concerned with the
question of climatic stability in the Canadian Arctic for

45

the past three centuries. In path-breaking work, they subjected the journals of Company posts situated on Hudson Bay to rigorous content analyses and they were able to demonstrate that river break-up and freeze-up dates had not changed significantly during the historic period.[2] These findings strongly suggest that the length and severity of the winter season has not varied substantially since the late seventeenth century.[3]

Recently Moodie has pointed out that a wide variety of other environmental data can be extracted from the Company's archives.[4] Included is detailed information regarding flora, fauna, and soil resources. Given that Moodie and Catchpole have already pointed the way, the potential of the Hudson's Bay Company Archives for historical environmental research will not be discussed any further. Instead, the prospects for work in medical, nutritional, technological, and architectural history will be explored.

The question of the impact that European diseases had on native populations has occupied the attention of anthropologists and geographers for a long time.[5] However, one of the vexing problems that frequently plagues those who pursue this line of enquiry is that data regarding the frequency, paths of diffusion, and severity of epidemics is fragmentary in many areas. Fortunately, Canada is an exception in this regard. The nature of the Hudson's Bay Company's trading network, the Company's system of record keeping, and the completeness of the surviving manuscript collection makes it possible to study the outbreak and spread of epidemics amongst Indian peoples in a degree of detail unattainable in most other areas of North America. In the nineteenth century, the Hudson's Bay Company operated an extensive network of posts that covered Canada from coast to coast. After 1821, operations were divided into departments (Montreal, Southern, and Northern) and these in turn were broken down into districts. The various departments and districts were supplied during the summer by a variety of brigades of boats, canoes, cart trains, and other craft. Contact was maintained in the winter as well

by the so-called "winter express" which carried correspondence.

Of importance to historical epidemiological research,
the Company's system of posts can be regarded as a network
of observation stations. And, its transport brigades may
be viewed as potential vectors in the spread of disease.
The primary problem that the researcher faces in attempting
to put these facts to work relates to the potential volume
of archival material that might have to be examined before
the frequency and patterns of the epidemics could be established. For instance, the Hudson's Bay Company Archives
include records for at least 239 different posts. Granting
that most of these records are for the period after 1774,
and further allowing for the fact that collections of manuscripts for many posts are not complete, one would still
face a formidable task if an attempt were to be made to
read all of the post journals. Fortunately this is not
necessary, especially if one is dealing with the period
after 1821.

In the early years of the Company's operations, the
men in charge of the posts (the Chief Traders or Factors)
were expected to file an annual report to the Governor and
Committee in London, England. In these reports they were
supposed to comment on the state of the trade. Needless
to say, poor returns called for an explanation. Most
commonly, a scarcity of game, the outbreak of disease, or
warfare were cited as the causes. Thus, by examining the
inward correspondence to the Governor and Committee it is
usually possible to learn whether or not any significant
epidemics occurred in a given year.

After 1821, a governor was placed in charge of the
Company's Canadian operations. The first of these governors
was Sir George Simpson. The factors and chief traders
generally wrote two official letters to the governor each
year; one was sent via the "winter express" and the other
was dispatched with the returning brigades in the summer.
In addition, these men were expected to file annual district reports. For these reasons, one can obtain a good
overview of the state of the fur trade after 1821 by

examining the inward correspondence of governors who re-
sided in North America and the district reports.[6] The
latter documents are particularly useful for the period
between 1821 and 1830, and the former for the period
thereafter.[7] By proceeding in this fashion, one can
nearly always pinpoint the districts and posts where significant illness occurred.

Analyses can be further refined by examining the
relevant post journals. By doing so, very often it is
possible to pinpoint the day, and frequently the hour, when
a given epidemic first manifested itself locally. Furthermore, these journals enable one to track the progress of
the disease in the environs of the post and to assess its
impact on the local population. The utility of conducting
research in this fashion has already been demonstrated. By
examining the records for the Northern Department of the
Hudson's Bay Company for the period 1830-1850, the author
was able to map the spread of six major outbreaks of in-
fluenza, and one each of scarlet fever, measles, and
smallpox.[8] It was also possible to identify the vectors
for each epidemic and the factors that favoured, or hin-
dered, the transmission of disease to the native populations.

Besides the area of epidemiology, the Hudson's Bay
Archives offer other avenues of research for the historian
of Canadian medicine. For example, many of the company's
officers were trained doctors. Probably the most famous
of these was Dr. John McLaughlin, who was in charge of
trading operations on the west coast in the nineteenth
century. In addition to medical doctors who acted as factors and traders, others were hired to serve as surgeons
at important posts or settlements such as York Factory and
Red River. Yet, to date, no one has studied the medical
careers of these men and assessed the efficacy of their
methods in light of the state of western and traditional
native medical practices of their day.

A variety of records exist that will enable scholars
to undertake such studies. Most obvious, of course, are
the letters and journals that these men kept in their

capacities as clerks and traders. Indeed, besides offering
information about their diagnoses and treatment of
diseases, these documents offer information about how these
men felt about their dual careers as doctors and traders.
One of them, Dr. William Todd, the most famous fur trade
doctor in the western interior between 1821 and 1850,
believed that he suffered a great deal in terms of physical
health and material well-being because he was a medical
doctor. Most of those who served dual roles believed that
they were overworked. Some medical records also exist.
At the more important posts, an account was sometimes kept
of diagnoses and prescribed treatment. For example, at
York Factory, such a record was kept in the form of a
medical journal. It covers the years from 1846 to 1852.[9]
Occasionally, the health of the men at a given post was
poor due either to local outbreaks of disease and problems
with water supplies and/or diet. At such times inquests
were held and reports filed. For instance, between 1833
and 1836, the men of York Factory suffered severe cases of
colic accompanied with vomiting, restlessness, severe pain,
and depression. The symptoms broke out in late winter and
in severe cases led to convulsions, loss of reason, and
death.[10] In 1837 a surgeon, Dr. Musgrove, filed a report.
He examined the water, food, and exercise habits of the
men at the post. He did not identify the ailment, but
indicated that he believed improper food preparation, a
lack of vegetables, insufficient exercises (especially by
the officers), and over-heated rooms were "predisposing
causes."[11]

One might suspect that the cause of this "York Factory Complaint" was scurvy, or some other dietary disorder,
but this is questionable since company men knew both scurvey's symptoms and treatment. Lemon juice had been imported
into the bay for some time but no other bay-side post reported either the "York Factory Complaint" or scurvy. Another incident in this episode suggests that dietary deficiencies are not the key. In the summer of 1836, York Factory's surgeon, Dr. Whiffen, was so reduced by the illness
he was sent inland to recover.[12] Dr. William Todd was his

replacement. Within one week of arriving at York Factory, Dr. Todd was struck by the same illness. He became so seriously ill that everyone thought that he would not survive. However, after ten weeks of what Dr. Todd described as suffering what was little short of torture, he recovered. He also claimed to have dealt with the source of the problem at York Factory and no further cases were reported.[13] Unfortunately, he did not indicate what measures he took, nor does he say what he thought the disease was. Hence, we are left with an unresolved riddle.

The incident at York Factory serves to indicate another area where fruitful research could be undertaken-- namely nutritional science history. While the "York Factory Complaint" episode does not appear to have had its roots in a vitamin deficiency, nonetheless, Dr. Whiffen and others indicate that scurvy was a recurrent problem at York Factory in the first half of the nineteenth century in spite of the fact that lemon juice was being imported to prevent it.[14] The enigma relates to the fact that the other bay-side posts do not appear to have been plagued by the same scourge. One wonders then whether the scurvy diagnosis was accurate? If it was, why did the other posts not experience it?

There is a wealth of information in the company's archives that could be brought to bear in studies of Indian and European nutritional habits and problems resulting therefrom. For example, for the period before 1763, the "State of the Stores" sections of the account books provide detailed information regarding the quantities and varieties of food that were being imported from Europe.[15] The stores accounts and the expenses sections of the account books also provide data regarding the provisions that were being obtained from the Indians. The early account books also occasionally contain mess books that give a detailed daily breakdown of the food being consumed at the post. One of the earliest such records is that for York Factory for 1727-28.[16] These kinds of documents are more common for the late eighteenth and early nineteenth centuries. They are variously titled Provision, or Provision Shed Blotter Books. The record

for York Factory is particularly rich. The provision shed
books provide a complete account of diet at York Factory
from 1827 to 1863 with the exception of a ten-year period
between 1832 and 1842.[17] However, this gap can be largely
filled in by examining the account book records for those
years.[18]

The provisioning records cited here are only intended
to be indicative of the kind of information that is available. It is hoped that they will serve to encourage those
who are interested in the history of nutrition to probe
more deeply. Fur trade and native history scholarship
would greatly benefit from studies that systematically
examined life-ways and work habits of Indians and traders,
determined their nutritional needs, and related these requirements to their actual diets. Needless to say, such
work should be an essential part of any meaningful studies
of cultural ecology. Yet, to date, no work of this type
has been undertaken.

Besides medical and nutritional history, technological
adaptation and innovation is another subject that can be
fruitfully explored in the Hudson's Bay Company Archives.
Until recently, little thought was given to the process
whereby Indians adopted European tools and technology. Of
particular importance, no consideration has been given to
the role that the Indian might have played in forcing
Europeans to modify and improve the articles that they
offered in trade. Recently it has been shown Indians did
not accept European firearms, metal tools, and utensils
without comment. Rather, being astute traders and consum-
ers, they had well-defined tastes and expected trade goods
to meet certain standards.[19]

Europeans had considerable difficulty meeting these
demands. The problem appears to have been twofold. In
abandoning his stone age technology for one of the iron
age, the Indian also was placing himself in a more vulner-
able position. Whereas stone, bone, and wood tools were
not as durable as iron, nonetheless, broken implements
could be easily repaired or replaced with available technology and materials. This was no longer the case when

ironware and firearms were substituted for them. Thus, European goods had to be durable and well suited to the nomadic existence of the Indian who lived in the harsh subarctic environment. The Europeans, on the other hand, initially sold tools that had been designed for use in Europe, or for trade in tropical and subtropical regions such as India.[20] Given the requirements of the Indians, and the environments for which European goods had been developed, it is not surprising that problems arose.

Most of the difficulties centred around articles that were manufactured from iron. Problems appear to have involved both design and technological capability. Regarding design, the nomadic lifestyle of the Indian dictated that goods be as lightweight as was possible and yet able to withstand the severe cold of the winter. This was a tall order and the record suggests that it taxed the limits of European technology in the eighteenth century. For example, the Indians frequently complained that metal kettles broke. The Hudson's Bay Company reacted to these criticisms by making them thicker. This did not solve the problem, however, because the Indians then judged them to be too heavy.[21] Firearms posed a similar problem. The Indians wanted lightweight arms. In an effort to meet this demand, the Company ordered its gunsmiths to file the barrels thinner. At that point the Indians complained that the barrels were bursting all too frequently. The exasperated London directors then told their traders in the bay to inform the Indians that they would have to decide what was more important, weight or durability.[22] Apparently the English gunsmiths could not satisfactorily meet both requirements.

Casting, soldering, and welding techniques were also a source of trouble. For example, the Indians frequently would not accept firearms and other metal tools because of blemishes that they observed in the metal. This infuriated the Governor and Committee in London. They claimed that these imperfections were "fire flaws" and that arms and metal tools could not be made without them. Indeed, the London directors argued that such flaws were proof of the hardness of the metal and would not cause the metal to fail.[23] In spite of these claims, the Indians refused to accept many goods that had these pin holes and the traders were forced to return them to England.

Places where pieces of metal were joined together proved to be sources of trouble as well. All too often handles broke off of kettles, axe heads split, and so forth. Indians quickly learned where metal tools were likely to fail and were said to be able to detect even the most minute cracks in joints.

The picture that emerges from a quick reading of the record suggests that a technologically inferior people, the Indians, were placing demands on a technologically superior people, the Europeans, that the latter were unable to meet. English craftsmen had had no prior experience in designing tools for arctic and subarctic conditions before 1670. The Company attempted to deal with this problem in two ways. They instructed their traders in Hudson's Bay to find out exactly what the Indians disliked about English goods. Defective merchandise was returned to England with the flaws carefully marked. The Governor and Committee in London passed this information on to their manufacturers and suppliers requesting them to rectify the problems. Also, they bought samples of French goods from the Indians whenever the latter expressed a preference for them and had them copied in England.

Besides pursuing this course of action, the Company also decided to manufacture some of its metal wares and other goods at its posts on Hudson Bay. This idea had been suggested by Governor John Nixon within ten years of the founding of the Company in 1670. Nixon had despaired of ever obtaining suitable merchandise in England and suggested the Company would be better off if it sent blacksmiths and other craftsmen, along with the required raw materials, to the bay. There, he argued, the Company's goods could be manufactured more to the Indians' satisfaction. Eventually many goods were made at the bay. Indeed, some were made exclusively there. For example, by 1763, all of the ice chisels that were sold at Fort Churchill

were fashioned there. According to the Company's traders,
the Indians preferred those that the local blacksmith made
to any that could be purchased in England.[24] Perhaps the
Fort Churchill blacksmiths had an advantage over their
English counterparts in that they could experiment and
test their work in the environment in which it would
ultimately be used.

In any event, this discussion of technological innovation and adaptation is intended only to indicate the kinds
of questions that need to be dealt with by competent historians of technology. What were the technological capa-
bilities of European blacksmiths and tinsmiths in the
eighteenth century? Was metallurgy and smithing sufficiently advanced to deal with arctic conditions? Did the
fur trade lead to technological innovations? What kind of
advantages did locally based craftsmen have over their
European counterparts? How much manufacturing was done
at the posts, and elsewhere in Canada, to meet the demands
of the fur trade? To date little research has been done
that deals with these issues. In large part, this is no
doubt a reflection of the fact that fur trade and native
historians do not have the technical expertise to deal
with such questions in any depth.

For those scholars who do have such a background, the
archives of the Hudson's Bay Company are exceptionally
rich. For example, the Official London Correspondence
Inward and Outward deals extensively with the problem of
obtaining suitable trading goods. These documents cover
the entire period from 1670 to 1870. The Minute Books,
Invoice Books, Grand Journals, and Grand Ledgers enable one
to determine who was supplying merchandise to the Company
and in what quantity.[25] The correspondence books cited
above provide a commentary on the quality. Concerning
manufacturing activities in Canada, the account books are
rich sources. For example, in the post account books are
included data regarding the state of the blacksmith,
armourer (gunsmith), carpenter, and other "stores." These
statements provide detailed inventories of the tools that
are on hand, those that were broken or discarded, and the

supplies that were used in manufacture or repair.[26]

By the nineteenth century, extensive manufacturing was being done at major posts in addition to an ever increasing amount of repair work. Reflecting this, the stores accounts are supplemented by a much wider array of information. For example, the York Factory records contain what are called tradesmen's work lists. Here under the headings of blacksmiths, carpenters, coopers, painters, sailmakers, tinsmiths, women, kitchen men, invalids, and storesmen are the lists of work to be done. These lists indicate the quantities of a given item needed on inventory for the year, the total presently on hand, and the number that need to be made at the post.

In addition, information is often provided regarding the operations of a particular shop. For instance, under an account entitled "Blacksmith Forge, 1834-40" one can determine the materials used, the number of men, and the days of work that were required to produce a given number of articles. It is laid out in a debit-credit format. On the left-hand debit page, materials, labourers, and hours of work are entered, while on the right-hand, or credit, page, the items produced were listed.[27] Cost estimates of this production are provided in other documents.[28]

Once again, no attempt has been made to detail all of the extant records that are pertinent to the study of Hudson's Bay Company manufacturing in Canada. The intent has been merely to indicate the richness and variety of material that is available and some of the questions that need to be answered.

Finally, it should be pointed out that the Hudson's Bay Company's struggle to cope with the realities of the arctic and subarctic environment involved adjusting western European architectural practices so that structures could be built for conditions unlike those found anywhere in the British Isles. The depot at York Factory is a testament to the skill of the Company men who met that challenge. The York Factory depot is a wooden structure that was built in the 1830s. The builders had to contend with poor drainage conditions and permafrost. They dealt with these

problems by erecting the building over a drainage ditch.
The structure is supported by corner sills that rest on
pilings sunk two feet into the ground. To deal with frost
heaving, the ground floor is laid over runners placed on
the ground. It is not attached to the walls so that it
can move independently of them. Also, the beams and internal supporting posts that hold up the second floor are
joined in a manner that allows workmen to make the adjust-
ments that might be required due to frost heaving or the
settling of the building. In spite of the fact that the
building is over 150 years old, it is in remarkably good
condition. Permafrost has not ravaged it the way it has
many other buildings of more recent vintage, such as those
of Dawson. Yet, to date, no architectural history has been
done of the York Factory depot. One is urgently needed.

In conclusion, an attempt has been made to indicate
some of the areas where research needs to be done in the
field of the history of Canadian science and technology.
From the beginning, the development of Canada, in a large
measure, involved learning how to deal with the arctic and
subarctic environments which encompass most of the country.
As early as the seventeenth century, science and technology
were keys in the struggle to meet this challenge. This
important story remains to be told. Fortunately, for any
who choose to head out on this research frontier, the
archives of the Hudson's Bay Company offer an exceptionally
rich and largely untapped mine of information. All the
researcher requires is the patience, imagination, and
ingenuity to formulate meaningful questions and systematically search the archives for answers.

NOTES

The author would like to thank the Hudson's Bay Company for
permission to consult and quote from the Company archives.
I would like to thank Mrs. Shirlee Anne Smith, Archivist,
Hudson's Bay Company Archives, Public Archives of Manitoba,
Winnipeg, Manitoba, for commenting on earlier drafts of
this paper.

1. Public Archives of Canada, *Manuscript Group 20*, Ottawa,
    1974: 168-9.

2. D. W. Moodie and A. J. W. Catchpole, *Environmental Data from Historical Documents by Content Analysis: Freeze-up and Break-up of Estuaries on Hudson Bay, 1714-1871* (Winnipeg: Department of Geography, University of Manitoba, 1975).

3. *Ibid.*

4. See D. W. Moodie, "The Hudson's Bay Company's Archives: A Resource for Historical Geography," *The Canadian Geographer* 21 (1978), 268-74; and A. J. Catchpole and D. W. Moodie, "Archives and the Environmental Scientist," *Archivaria* 6, 1978: 113-136.

5. See Henry F. Dobyns, *Native American Historical Demography: A Critical Bibliography.* (Bloomington: Indiana University Press for the Newberry Library, 1976) and William M. Denevan, ed., *The Native Population of the Americas in 1492* (Madison: University of Wisconsin Press, 1976).

6. The correspondence to the governors can be found in section D of the Hudson's Bay Company Archives. The Inward and Outward Correspondence of Governor Sir George Simpson covers the period between 1821 and 1860 and comprises 179 volumes (123 reels of microfilm) of letters. The District Reports can be found in the post records (Section B, subsection e).

7. In the early years, the traders filed reasonably complete reports. However, since much of the information was repeated year after year, the authors of the reports appear to have lost interest in the exercise, and as a consequence, the district reports become rather cryptic in later years.

8. Arthur J. Ray, "Diffusion of Diseases in the Western Interior of Canada, 1830-1850," *Geographical Review* 66: 2 (1976), 140-157.

9. York Factory Post Journals, 1846-47, 1847-48, 1848-49, 1849-50, 1850-51, 1851-52, B239/a/166, 169, 172, 174, 177 and 180.

10. Governor George Simpson, Letters Inward, 1834, E. Smith to Alexander Christie, 26 May 1834, D 4/126, 65-6.

11. York Factory, Miscellaneous, 1837, B 239/Z/26, 143-4.

12. Norway House Post Journal, 1836-37, B 154/a/27, 19.

13. London, Correspondence Inward, From Dr. William Todd, Norway House, 15 July 1849, A 11/51, 4-7.

14. For example, the York Factory Journal of 13 January 1836 contains the following entry: "It is most pleasing to have to add, that not a single case of Sickness at present exists in the Factory, and that even the lightest symptoms of Scurvey have not yet

shewn themselves among the People throughout the whole winter. This fact would seem to prove that the use of Salt provisions can scarcely be the principal cause of that disease at this place as the present Rations consist of as large a proportion of that food as usual say 3 days in the week." B 239/A/149, 23. This observation is instructive in that it indicates the journalist expected the disease to make its appearance. Also, he indicates that many of the men blamed the frequent outbreaks of scurvy on the diet at the post. In fact, many of them refused to eat the salt provisions and returned them.

15. See Arthur J. Ray, "The Early Hudson's Bay Company Account Books as Sources for Historical Research: An Analysis and Assessment," *Archivaria* 1 (Winter 1975-76), 3-38.

16. York Factory Account Book, 1727-28, B 239/d/18, 29d-36d.

17. Provision Shed Balance Books, York Factory, 1827-71, B 239/s/1-26.

18. York Factory Account Books, 1832-42, B 239/d/444, 446, 458, 459, 473, 476b, 480, 495b, 499, 515, 517, 518, 537, 538, 553, 556, 573, 601, 615, 619, and 640. In all, there are 21 volumes of material that deal with provisions during this period.

19. Arthur J. Ray, "Give Us Good Goods: The Indian as a Consumer in the 18th Century," Paper Presented at the 3rd North American Fur Trades Studies Conference, Winnipeg, May 1978.

20. *Ibid.* and Arthur J. Ray, "Why Study the Fur Trade: Trade History as an Aspect of Native History," in *One Century Later: Papers of the 9th Annual Western Canadian Studies Conference* (Vancouver: University of British Columbia Press, forthcoming).

21. Ray, "Give Us Good Goods."

22. *Ibid.*

23. *Ibid.*

24. *Ibid.*

25. The Minute Books provide information on discussions and decisions of the Governor and Committee regarding selection and suitability of suppliers. The Invoice Books indicate the quantities of goods being shipped from London and the post to which they are being sent. The Grand Journals and Grand Ledgers provide information regarding the names of suppliers, the quantities purchased from them, the unit prices paid and the aggregate value of purchases.

26. Ray, "Analysis and Assessment."

27. York Factory, Blacksmith's Work Book, 1834-40, B 239/
     d/568.

28. York Factory, Estimated Cost of Made Up Works,
     B 239/d/589.

EARLY FEDERAL REGULATORY RECORDS
AS POTENTIAL SOURCES FOR THE
HISTORY OF SCIENCE AND TECHNOLOGY IN CANADA:
THE CASE OF THE SAWDUST POLLUTION FILES, 1866-1902
Peter Gillis

During the last half of the nineteenth century various
levels of government in Canada involved themselves in a
number of activities not normally associated with the idea
of the negative state or the tenets of laissez faire.
This trend has been particularly noted in the field of
social welfare, public health, and utility regulation which
were primarily provincial and municipal responsibilities.
However, a similar process was also occurring at the federal
level. It has been a long-established thesis that the
new Dominion government continued, after 1867, and indeed
intensified the efforts of its predecessors in British North
America, to ensure economic development through the provision of transportation facilities and the promotion of
immigration.[1] Sir John A. Macdonald enshrined this system
as his "National Policy," adding the protective tariff to
his plans for a transcontinental railway and the settlement
of the North-West.[2]

This development policy, which remained relatively
unaltered by Wilfrid Laurier and his Liberals when they
took office in 1896, required a heavier government superstructure to ensure its success. Settlement and exploitation of resources created the need for an efficient and
permanent system of government surveys and land division,
for law enforcement, for supervision of immigration, for
experimental farms, and for greater efforts by the Geological Survey. By 1900 its development role in western Canada
was carrying the federal government beyond the simple ethic
of exploitation to one of resource management. Forestry
programs, irrigation work, regulation of water-power, wildlife conservation, and national parks all grew to be important aspects of Dominion policy during the first two decades

60

of the twentieth century. These, however, were not the
only new areas of concern for federal authorities. Increased industrial growth and urbanization after 1870 made
obvious a host of social abuses and the appearance of
large, integrated business corporations transformed the
economic structure of the nation. Prompted by public
pressures and often directed by bureaucratic initiative
the federal government reacted in an ad hoc manner to
demands for new regulatory legislation. Slowly but surely
it moved into a supervisory role over railways and telephone companies, public health, pure food, and labour relations among others.

Historians in Canada are only beginning to probe these
various activities and ferret out the multifaceted origins
of the positive state in this country.[3] Few scholars,
however, have attempted to use the files created by early
regulatory agencies for the study of the history of
science and technology. This is unfortunate because these
records contain a great deal of valuable information. The
growth of regulatory legislation at the federal level after
1870 meant that a greater number of civil servants had to
be hired to administer the new acts. But enforcement of
statutes and the monitoring of conditions in industry often
required much more expertise than in the past. A cursory
glance at the *Civil Service Lists* from 1878 until 1916
reveals an ever increasing number of professional men
being attracted into government service.[4] Engineers had
found a home in government departments at an early date
but now they came in far greater numbers along with
scientists, doctors, and veterinarians, as well as several
other categories of professionals. They were an entirely
new phenomenon, a modern breed of civil servant who had an
important role to fulfill within the expanding functions
of the state. A very few, particularly in the Department
of Agriculture, were involved in pure research, but most
performed investigatory and regulatory work where their
knowledge was applied in a highly practical way to inspect
various industrial activities and products, and to
evaluate a host of projects, both public and private. These

employees left behind them a wide variety of records.
Many files are comprised only of inspection forms which
provide the researcher with very little information.[5]
Others, however, especially those relating to the first
implementation of new regulations, contain engineering
reports, plans, descriptions of industrial structures and
processes, reports from scientists on problems and possible
solutions, research notes, and correspondence from those
against whom the regulations were being enforced, giving
reasons for non-compliance or information on the modifications being made. Of course it is impossible in a short
paper to cover every type of regulatory record. Instead
I have chosen to deal with one series of files which document an early government campaign against water pollution.
Involved is the work of Department of Marine and Fisheries
personnel to end the dumping of sawdust from lumber mills
in Canadian rivers and the variety of source materials
available on these files provides an excellent illustra-
tion of the rich potential of such early regulatory records
for the history of science and technology in Canada. Indeed
the sawdust pollution files aptly demonstrate both the important nature of the information found in such early series
of records and also the impact of the shift from the older
civil service to the new which made the collection of such
documentation possible.

I

Sawdust pollution was one of the great environmental
problems of the late nineteenth century in Canada. The
lumber trade was the nation's leading nonagricultural industry and provided an essential export commodity.[6] International markets had been developed in Great Britain, the
United States, and South America. The country's prosperity
was largely dependent on the health of the lumber industry
and from Nova Scotia to the new North-West hundreds of
lumber mills churned out the milky pine so prized in construction around the world. The vast majority of these
plants were water-powered and the edgings, sawdust, and
other mill refuse produced by the saws was simply dumped
in the streams which bordered the mills.

The effect of this wanton pollution was immediately
apparent to those living along the rivers of Canada. Sawdust, floating downstream in large patches, settled in
bays, fouling spawning beds for fish and destroying
much of the insect life on which they fed; then it began
to move out into the main channel to obstruct navigation;
and finally, mixed with sewage, it began to rot, creating
obnoxious gases which were prone to explode through spon-
taneous combustion and which rendered the water impotable.[7]
The first definite legislation against this nuisance was
passed in the Province of Canada in 1865 as part of "An
Act . . . for the Better Regulation of Fishing and Protection of Fisheries."[8] Pushed on by the Officer of Health
for the City of Ottawa, Dr. E. Van Courtland, the Department
of Crown Lands, which was responsible for enforcing the
act, warned mill owners that they "must adapt their
premises to the disposal of waste materials in such a manner
as shall obviate further injury to the rivers and streams."[9]
This confrontation in 1866 between fisheries officers and
the mill owners was the first salvo in a battle which was
to endure of the remainder of the nineteenth century.
Centred on the mills around the Chaudière Falls in Ottawa,
where some of the largest mills in Canada were located, it
was watched intently by lumbermen in Nova Scotia, New
Brunswick, and Québec because, if the Ottawa mills were
forced to comply, then they too would have to follow. It
was a classical fight between an industry which feared that
the installation of antipollution machinery would prove too
expensive and ruin the competitive position of those companies which were forced to modify their plants and civil
servants charged with protecting inland fisheries on navigable streams, navigation on those streams, and general
public health and welfare. Political interference was
rife both because of the power of the lumber interests and
because the operators threatened to close down their mills,
throwing thousands out of work.[10] On the other hand there
was a vocal antipollution group centred in the Senate which
kept the issue before the public and put pressure on fisheries officers to take strong action against the

lumbermen.[11] Only amendments to the Fisheries Act in
1894, which contained strong clauses against sawdust pollution, excluding exemptions of all kinds, and determined
efforts to enforce these regulations, which eventually
involved the intervention of Sir Wilfrid Laurier himself,
finally ended the abuse in 1902.[12]

In an archival sense this long controversy generated
an excellent variety of information beyond the straight
endeavour to enforce fisheries regulations. Mill technology is discussed at length by engineers and construction experts and, as well, a good deal of information is
provided on methods of determining water quality and the
deterioration of fish species. Finally, the files themselves reflect the evolution from the older type of civil
servant to the new, with the consequent expansion of the
kinds of documentation which find their way into the
record.

In a typical mid-Victorian approach the Honourable
Archibald Campbell, Minister of Crown Lands in 1866,
appointed an engineer to look into the matter of sawdust
pollution at the Chaudière. This decision reflected the
lack of desire on the part of the Department to strictly
enforce the new regulations. Facing vociferous petitions
from the lumber barons in Ottawa, Campbell wished to find
some practical and mutually agreeable solution to the
question.[13] The man he chose to investigate the operations
of the Ottawa mills was admirably equipped to do just that.
Horace Merrill was a self-made engineer who was Super-
intendent of the Ottawa River Works, within the Department
of Public Works. An imaginative construction designer, he
had long been friendly with the lumbermen in the Ottawa
Valley, where he had constructed slides and other river
improvements and was also a major partner in the Victoria
Foundry and Machine Shops, which specialized in the manufacture of mill equipment.[14]

Merrill had his obvious biases. He lamented the effect
sawdust had on the fish population in the lower Ottawa River
but quickly dismissed this as a question of no real consequence. Instead Merrill concentrated on the question of

navigation. This he found adversely affected by the
larger edgings and slabs which posed a threat to boat
traffic but not by the buildup of sawdust. The sole
exception to this general situation was at the mouth of
the Rideau Canal, which was becoming clogged.[15] Taking
this as the basis of his report he then went on to investigate and report on each of the mills in the valley below
the Chaudière as to the feasibility of installing grinders
to reduce all the waste to sawdust before it was dumped
in the river. Merrill then went on to design a grinder
which would suit these purposes and arranged for the
Victoria Foundry to produce it. By 1867 this plan had been
accepted by the Department of Crown Lands and Merrill was
ready to turn a good profit as a result of the recommendations of his report.[16]

Merrill's work was to set the trend for the work of
the new Department of Marine and Fisheries which became
responsible for enforcing the sawdust regulations after
1 July 1867. Continuing to take a strictly practical
approach to the situation, its officers tried to find
solutions which were not too expensive or complicated and,
therefore, would find acceptance with the lumbermen. Thus
between 1866 and 1877 Merrill, and later John Mather, a
former longtime manager of the Gilmour mills on the
Gatineau, amassed considerable documentation on the operations of the plants in Ottawa, reporting on their construction, the machinery in them, current mill technology and
how it could be applied to the problem, and finally on the
growth of steam-powered lumber plants; their advantage in
using sawdust for fuel and their disadvantage in creating
a smoke nuisance in an urban setting.[17]

Mather, a knowledgeable mill manager turned civil
servant, actually went much beyond Merrill's original reports on the mills. A new act entitled "An Act for the
Better Protection of Navigable Streams and Rivers" was
passed by Parliament in 1873.[18] It stipulated that no
sawdust or mill offal could be dumped in navigable streams
but provided for exemptions where it could be proven that
such regulations would adversely affect the lumber industry

Applications by a number of mill owners in Ottawa for exemption under the act persuaded the Department of Marine and Fisheries to have Mather prepare a detailed report on the problem in the capital. He spent nearly six months studying the disposal of mill waste in the lower Ottawa Valley and faithfully describes activities in most of the mills. He was also the first fisheries officer to begin surveys of the river, taking both soundings and water samples.[19] Estimating that 12,300 cubic feet of sawdust went into the Ottawa River annually, Mather declared the situation intolerable and, based on his detailed examinations of the mills, called for conveyors and tramways which would carry the waste to furnaces. His report stands as an eloquent status report on the plants themselves and the condition of the Ottawa River in 1878-79.[20] John Mather, however, despite his definite stand against the abuse of sawdust pollution, was only successful in getting the lumbermen to agree to the very same conditions Horace Merrill had negotiated from them. They agreed to be more thorough in chipping slabs and edgings but insisted on their right to dump sawdust in the river. Their official exemptions from the Act were proclaimed by an Order in Council in 1880.[21]

Complaints against the pollution practised by the mill owners continued unabated, however, and in late 1887 the Senate conducted a full-scale enquiry into the problem. Its report, which appeared on 15 May 1888, fully reopened the question and prompted demands for stricter regulations. Frightened by this turn of events, the lumbermen in Ottawa hired Sir Sandford Fleming to investigate and report on conditions in the Ottawa River.[22] Predictably Fleming found no great difficulties ensuing from sawdust dumping except at the mouth of the Rideau Canal, where it blocked navigation. He suggested that the mill owners undertake dredging at that point and have done with the whole question.[23]

The Department of Marine and Fisheries was not to be that easily bilked. The effect of a newer type of civil servant was beginning to make itself felt. Sensing that it might be possible to secure legislation which would

effectively prevent the dumping of mill offal in all
navigable rivers, its officers moved quickly to document
their case against the mill owners. The Department of
Public Works was approached for its Assistant Chief Engineer, Mr. Henry A. Gray, to undertake a detailed survey of
the Ottawa River from the Chaudière to Grenville in order
to determine the effects of sawdust deposits on it. In
contradiction to Fleming he found it a definite hazard to
navigation and tremendously damaging to the river itself.[24]
Further, Marine and Fisheries obtained the services of
Dr. W. A. McGill, Ph.D., F.R.S.C., Assistant Analyst for
the Department of Inland Revenue, to test the waters of
the river to find out whether it would propagate morbific
bacteria. It was found that it would and further tests
by other scientists, under contract to Marine and Fisheries,
proved that the City of Ottawa had one of the worst supplies of drinking water in the Dominion, caused largely
from the dumping of raw sewage in the Ottawa River but not
improved by the sawdust.[25]

Finally the Department of Marine and Fisheries used
one of their own experts to amass evidence of the bad
effects of sawdust pollution. He was Edward E. Prince,
a graduate of St. Andrews, Edinburgh, and Cambridge and a
specialist in fish food supply and spawning habits. He
joined the Canadian government in 1892 where he was to have
a very distinguished career in marine biology and as a member of several international boards and councils dealing
with fisheries.[26] His reports on the sawdust problem established beyond a doubt that this type of pollution had a
very detrimental effect on the fish population of rivers
and streams.

Thus by 1894 when the Fisheries Act was amended to
prohibit the dumping of mill waste in waterways the Department of Marine and Fisheries had come a long way in buttres-
sing its case against this abuse. Having done so its officers did far more than perform yeoman service in taking
action against a genuine environmental problem. At the
same time, by taking a broad approach to this problem,
they greatly enriched the archival sources for the study

of science and technology in Canada. The sawdust pollution
files provide a running commentary from 1866 until 1902
on mill structures and technology in some of the largest
lumber plants in the country. Further, they include some
daily activity sheets, plans, engineering reports for
modifications, and comparisons between water-powered and
steam-powered mills. As well, after 1894, there is
documentation on the files about experiments by the mill
owners to employ new technologies to mill waste to produce
pulp, fibre-board, and acetylene. The new type of civil
servant went beyond mill technology, however, into the
field of water quality and fish research. Papers and notes
appear on the files documenting experiments, commenting
on public health problems, and substantiating claims of
environmental deterioration on the Ottawa and other rivers.

This short descriptive summary has, I hope, served to
introduce you to the various types of information which can
be found on regulatory files. Such public records can be
used for many purposes, depending on how a researcher is
prepared to approach the documentation--to tell the story
of the regulatory agency itself but also to look in depth
at the industrial process, problem (i.e., disease, pollution, etc.), or other activity being regulated. The sawdust pollution files are not an isolated type of record.
Similar documentation can be found in the registries of the
Department of Railways and Canals, Canadian Transport
Commission, Supervisory Mining Engineer, the Dominion
Coal Board, Department of Agriculture, and the Department
of National Health and Welfare, to mention only the most
obvious. There much information remains, open to the public, but unexploited by those interested in researching the
history of science and technology in Canada.

NOTES

1. The best presentation of this thesis in H. G. S. Aitken,
   "Defensive Expansion: The State and Economic Growth
   in Canada," in *The State and Economic Growth*, ed.
   H. G. S. Aitken, (New York, 1959), 79-114.

2. For a full explanation of the National Policy consult
   R. Craig Brown, *The National Policy 1878-1902* (Toronto,
   1968).

3. Two excellent new administrative studies of federal agencies are M. Zaslow, *Reading the Rocks: The Story of the Geological Survey of Canada, 1842-1972* (Toronto, 1975) and J. Foster, *Working for Wildlife: The Beginning of Wildlife Conservation in Canada* (Toronto, 1977). Two other fine pieces of scholarship are H. V. Nelles, *The Politics of Development: Forests, Mines and Hydro-electric Power in Ontario, 1849-1941* (Toronto, 1974) and R. B. Splane, *Social Welfare in Ontario, 1791-1893* (Toronto, 1957).

4. Canada, *Civil Service Lists, 1878-1916* (Ottawa, 1878-1916).

5. An example of such files are the Cheese Factory licensing and inspection files which were held by the Department of Agriculture. Besides location and production figures these files provide no other information on the development of this important industry.

6. The best single work on the eastern Canadian forest industry is the A. R. M. Lower, *The North American Assault on the Canadian Forest* (Toronto, 1938).

7. For a general overview of the sawdust problem see *Report of the Select Committee of the Senate Appointed to Enquire into a Report Upon the Extent and Effect Upon the Ottawa River, of the Deposit Therein of Sawdust and Other Refuse, Ottawa, 1888.*

8. Province of Canada, *Statutes of 1866*, 19 Victoria, Chapter 11.

9. Public Archives of Canada, Department of Fisheries (R. G. 23), file 1669, pt. 1.

10. *Ibid.*, "Memorial sent to Lord Monck, Governor General of Canada by the Lumbermen of the Chaudière, 8 September 1866."

11. Canada, Senate, Sessional Paper No. 43C, 1891, and Canada, Senate, *Debates*, 25 June 1891, "Debate on Sawdust Bill," and 1894, 720-35.

12. Fisheries, file 1669, pt. 3, Sir Wilfrid Laurier to F. Gourdeau, Deputy Minister of Fisheries, 24 September 1901.

13. *Ibid.*, pt. 1, J. M. Currier to Honourable A. Campbell, 12 September 1866.

14. See S. Gillis "Horace Merrill," unpublished manuscript, Ottawa, Canada.

15. Fisheries, file 1669, pt. 1, Report of Horace Merrill, 12 December 1866.

16. *Ibid.*, Horace Merrill to Honourable A. Campbell, 4 March 1867.

17. *Ibid.*, pt. 2, John Mather's Report, June 1877.

18. Canada, *Statutes, 1873*, 23 Victoria, Chapter 65.

19. Fisheries, file 1669, pt. 2, Mather's Report, June 1877.

20. *Ibid.*, and Appendix 3, *Report of the Commissioner of Fisheries, 1878.*

21. Fisheries, file 1669, pt. 1, Orders-in-Council, 23 June 1880.

22. Public Archives of Canada, Bronsons and Weston Lumber Company Records, volume 111, Fleming Report and Correspondence.

23. *Ibid.*

24. Fisheries, file 1669, pt. 2, Gray's Report.

25. *Ibid.*, McGill's Report and other scientific documentation.

26. *Ibid.*, Prince's Reports and Recommendations; and *The Canadian Men and Women of the Time: A Handbook of Canadian Biography of Living Characters*, ed. H. J. Morgan (Toronto, 1912), 918-19.

PROBLEMS IN THE STUDY OF THE HISTORY OF
CANADIAN SCIENCE AND TECHNOLOGY

III

L'HISTOIRE DES SCIENCES ET DE LA
TECHNOLOGIE CANADIENNES ET SES PROBLEMES

INTRODUCTION

The problems of historical research and writing are
not limited to theoretical issues, as the historiographical
articles in Part I discuss, but also encompass some practical problems. Three of the most important of these are:
how to acquire the money to undertake research, how to
publish the results, and, increasingly demanding our
attention, why are we doing this research? The three
papers in this section address themselves to these issues.

We believe, and M. Mougeot has supplied some evidence
for our belief, that there is no lack of funding for historical research in the history of Canadian science and
technology. If the field lacks anything, it is researchers
who can pose substantial problems and seek financial
support to underwrite their projects. Of course, in a
young field, we cannot expect to find many professional
historians at work, and the Social Science and Humanities
Research Council data underlines this. Perhaps in the
future, funding difficulties will grow acutely, but that
time is not now.

Publication of history raises different kinds of
problems. In his contribution, Mr. Montagnes discusses
the obstacles in the way of publishing and marketing historical monographs. Those which he enumerates are not
peculiar to this field, but to all scholarly publishing
in the Western world. For journal articles, which he does
not consider at length, but which form the bulk of professional historical output, the problem is not, as yet,
critical. There are stumbling blocks and delays, to be
sure, but we have seen no evidence that high-quality
articles do not eventually find appropriate outlets in
scholarly journals. Montagnes mentions the probability
that the form of publication may change radically in the
near future (e.g., to microforms or computer-storage); while
we would all find these formats difficult to adjust to, we
may be faced with no other alternatives as the volume of
research grows and the financial base shrinks.

72

INTRODUCTION

Les problèmes liés à la recherche et à l'énonciation historiques ne sont pas limités à des questions de théorie, comme le laissent croire les articles historiographiques de la première partie: il ne faut pas oublier les problèmes pratiques. Parmi les plus importants, on singularisera les trois suivants: comment trouver du financement pour la recherche? Comment publier les résultats de la recherche? Et de plus en plus fréquemment, on demande pourquoi nous faisons cette recherche. Les trois textes de la section qui suit tournent autour de ces questions.

Nous pensons, et M. Mougeot pourvoit quelques données favorables à notre thèse, que la recherche dans le domaine des sciences et des techniques canadiennes ne manque pas de sources de financement. Ce dont cette spécialité manque le plus, c'est de chercheurs capables de poser des questions importantes et de rechercher les moyens nécessaires pour soutenir financièrement leur projet. Il est entendu que dans une spécialité nouvelle, on ne peut s'attendre à rencontrer de nombreux professionels de l'histoire au travail, et les données émanant du CRSHC soulignent ce fait. Il est possible qu'à l'avenir les difficultés de financement augmentent énormément, mais ce n'est pas le cas actuellement.

La publication des études historiques se heurte à des problèmes différents. Dans sa communication, M. Montagnes analyse les obstacles qui s'opposent à l'impression et à la diffusion des monographies historiques. Ceux qu'il décrit ne sont pas spécifiques à l'histoire des sciences; au contraire, ils se retrouvent dans toute l'édition savante du monde occidental. Dans le cas d'articles pour revues savantes, cas qu'il n'analyse pas à fond bien qu'il constitue la majeure partie de la production des historiens, le problème n'a pas encore atteint des proportions alarmantes. Bien sûr, diffi-

73

Finally, Mr. Wynne-Edwards tackles some of the issues
of the use of science and technology by decision-makers
in particular and society in general. As historians, we
rarely give much thought to whether our work has a broader
significance to the public, but one can argue--and we
believe that we had better be prepared to argue--that the
history of Canadian science and technology is essential to
an understanding of the nature of Canadian science and
technology today and in the future. Those who have read
even a portion of the burgeoning science policy literature
in Canada have seen how these documents are relatively
weakened by the lack of historical perspective. Science
policy writers cannot fill this lacuna, only historians
can.

cultés et retards existent, mais nous n'avons pas de
preuve pour démontrer que des articles de haut niveau ne
trouvent pas tôt ou tard le moyen de se faire diffuser
adéquatement dans des revues savantes. Montagnes indique
que, de toute probabilité, l'édition va changer radicalement dans un avenir proche (p. ex. introduction de micro-
formats et stockage sur mémoire d'ordinateur). Nous au-
rons certaines difficultés à nous adapter à ces nouveaux
supports de l'information, mais il sera peut-être impossible de les éviter si la croissance de la recherche s'ac-
compagne d'une diminution de la base financière de l'édition.

Enfin, M. Wynne-Edwards s'attaque à quelques unes des
questions portant sur l'utilisation des sciences et des techniques par la société en général et ses dirigeants en particulier. Comme historiens, nous ne prêtons pas souvent attention au sens que nos travaux peuvent avoir aux yeux du public, mais on peut avancer l'argument--et, à notre avis nous
aurions intérêt â nous préparer dans ce sens--que l'histoire
des sciences et des techniques canadiennes constituera une
base essentielle pour saisir la nature essentielle des sciences et des techniques canadiennes aujourd'hui et à l'avenir. Ceux qui ont lu ne serait-ce qu'une fraction des écrits
qui se multiplient au sujet de la politique des sciences au
Canada perçoivent à quel point ces documents se trouvent
relativement affaiblis par manque d'une perspective historique. Les politiciens de la science ne peuvent combler cette
lacune par eux-mêmes: seuls les historiens peuvent le faire.

THE ROLE OF GRANTING AGENCIES IN
RESEARCH IN THE HISTORY OF SCIENCE AND TECHNOLOGY,
PAST AND PRESENT
Yves Mougeot

Given the subject of this conference, that is, the
study of the history of *Canadian* science and technology,
and of this panel which is entitled, "*Problems* in the
Study of the History of Canadian Science and Technology,"
I thought it would be appropriate to give you some indication of the support we have provided in recent years to
historians of science and technology. In compiling this
information I felt that it would help us understand what
are some of the major activities in your field, and that
it would allow us to evaluate whether, from our point of
view, that of administrators of a granting body, there
are specific problems relating to the supply-demand component of funding.

I should remind you, at this point, that the S.S.H.R.C.
(like the Canada Council previously), has established no
priorities or "privileged" areas of research in the
Humanities and Social Sciences. We have adopted an em-
pirical approach to research support. We have entertained
a vast number of projects with clearly stated scholarly
objectives that have fitted our terms of reference, and
have relied for decision-making on peer evaluation: the
basis for the adjudication of requests. In view of our
procedures, you will appreciate that we have not made constant or periodic assessments of our programs, since we
were assuming, and are presently assuming, that the
evaluators whom we consult on the scholarly significance
and contribution to knowledge of individual projects also
indirectly express their views on the needs of the discipline as a whole. Therefore we have been and are continuing to operate on the basis of individual assessments of
proposals, rather than of a global adjudication of our
objectives, procedures, and achievements. However, a

global evaluation of our programs is presently being carried out, and I shall return to this aspect later on.

Given the fact, as I just mentioned, that I cannot refer to any specific analysis of our achievements over past years as a granting body, it is difficult for me to identify or to pinpoint specific problems in the field of history or more particularly in the field of the history of science and technology. Nonetheless, I have attempted to examine past applications in the field of the history of science and technology with the purpose of identifying specific characteristics.

Although this may appear to be at first glance a simple task, I should like to emphasize that a number of problems arose, mainly in view of the absence of a sophisticated system of classification of applications at the Council. For instance, applications in the field of the history of science and technology are classified with history applications: thus, in order to identify projects in the area which concerns us, one has to look at every card under "History" and isolate what are believed to be topics in the history of science and technology. This, as you may already have guessed, is not a very reliable method since titles are very often misleading, and one would honestly have to refer at least to the summary of the project, and in many cases to the detailed description of the project, in order to isolate applications in the field of the history of science and technology. Moreover, to carry out any rigorous analysis it would be necessary to review cards filed under other disciplines. Here, for example, I can mention applications submitted by scholars working in departments of philosophy, and dealing with a number of related subjects concerning epistemology and philosophy of science. In any case, any in-depth analysis of applications submitted in your field would require a clear definition of precisely what constitutes a topic in the history of science and technology. I regret to admit that these important steps have not been taken, simply because of the amount of work that this would have required and because of a lack of time. Consequently, the few

observations I will make are not based on a rigorous
analysis of applications submitted to our Council. It is
but a quick survey, but one which nevertheless ought to
provide some indication of the research carried out in your
domain.

Since 1973, I have kept a record of applications that
have been assigned to me. In view of the fact that I am
responsible for all applications in the field of history
which, as I mentioned earlier, includes applications in
the field of the history of science and technology, I
believe that I have been able to trace the vast majority
of applications in your area. I have also inherited data
on this topic from my predecessor and have obtained from
my colleagues information about some of the files they
have processed, and which might be considered relevant.
To sum up, I believe that I was able to trace data for the
vast majority of dossiers in the field of the history of
science and technology, covering an eleven-year period from
1968 to 1978. It should be noted that I will be referring
(in almost all cases) to files identified by the applicants themselves as projects in the history of science or
the history of technology; only a small number of applicants listed their project under another discipline. You
should therefore keep in mind that this is not an exhaus-
tive treatment of the subject and that it is rather a
sample than anything else, but a sample which, I believe,
covers most of the projects and the most important projects
submitted to the Council at the Research Grants Program in
the history of science and technology. The few figures I
will quote provide a basis of comparison which to some
extent surely squares with reality.

Now turning to specific questions: first, the number
of applications submitted to the Council over a period of
eleven years from 1968 to 1978 adds up to 58. I must
mention that I have not included those applications which
were withdrawn and those which are presently being entertained. As you can see, this gives an average of 5.3
applications per year. Now looking at the details, we
arrive at the following figures:

Year	Requests in the history of science & technology
1968	2
1969	5
1970	5
1971	6
1972	4
1973	7
1974	7
1975	10
1976	3
1977	5
1978	4

We reached a peak during the 1973-1974-1975 period: 24 projects or 41 percent of all applications were submitted during those three years. The remaining applications (34), which make up 59 percent of the total, were submitted during a seven-year period. The major observation to be made here is that after 1975 there is a sharp drop.

How does this compare with the general situation with regard to history applications? The following figures are given only to provide some idea of the number of projects we have received and cannot be compared with the numbers quoted above since the method of compilation differs. The period is also shorter.

Year[1]	Requests in history
1972-1973	195
1973-1974	174
1974-1975	202
1975-1976	156
1976-1977	125

Here again, we seem to have reached a peak in 1975 and are witnessing a decrease in requests after that year; in fact, the decline has been constant since that time. It appears, therefore, that scholars in the field of history of science and technology are following the pattern of their colleagues in other areas of history.

Another question of interest is the success rate. Do we appear to have a problem at this level? For instance,

have we discouraged scholars from applying by establishing
very tough criteria resulting in a very low success rate?
In view of the results obtained by specialists in your
field, the situation with respect to the success rate
appears, on the contrary, very encouraging. In fact, the
success rate for requests in the history of science and
technology over an eleven-year period is 84 percent. Again,
how does this compare with the general success rate at
the Council?

Success rate of applications

Year	In history	All disciplines combined
1972-1973	88%	77%
1973-1974	82%	74%
1974-1975	78%	68%
1975-1976	85%	69%
1976-1977	76%	70%
Average success rate over a five-year period:	81.8%	71.6%

Given that the success rate of applicants in the
field of history of science and technology is 84.4 percent
over the period 1968-1978 (or 80 percent for the period
1972-1977), it would appear that historians of science have
had little difficulty in getting support from the Council.
Their success rate compares favourably with the success
rate of other historians and it is definitely higher than
the general rate of all disciplines combined.

If, at first glance, the figure obtained for success
rates appears quite encouraging, it should not divert our
attention from the negative side of some of the statistics
quoted above. I am referring, of course, to the number
of scholars who request support from our Council, a number
which is relatively small. Approximately 58 proposals over
a period of eleven years is not something to rejoice about
particularly. In addition, although I mentioned 58 applications, we are in reality talking of only 32 historians
since some of these applicants have submitted requests for
renewals or have requested support for more than one topic.

Thirty-two applicants, therefore, over a period of eleven years, make an average of 2.9 applicants per year in the field of the history of science and technology.

Now, a few words about the amount awarded in order to give you a basis of comparison. During this eleven-year period, the applicants who were successful in obtaining support were awarded $353,210. If I isolate the years 1972 to 1977, the total comes to $246,899. According to the figures provided by our Research and Analysis Section, the Council has awarded $3,503,256 for projects in history and $25,325,567 for requests in all disciplines combined during the period 1972/1973 to 1976/1977. Historians of science received approximately 7 percent of the funds awarded for history and .9 percent of the total funds awarded for that period in research grants.

It is to be noted that the vast majority of scholars in history, including history of science and technology, usually request small amounts not exceeding $5,000.

Categories	Awards	Percentage
$1,000 - 5,000	28	58
5,001 - 10,000	11	23
10,001 - 25,000	8	17
25,001 -	1	2
TOTAL:	48	100

In fact, 58 percent of scholars in your field were awarded grants not exceeding $5,000 while 23 percent were in the upper category of $5,001 to $10,000. It is clear that the vast majority of historians of science and technology (81 percent) carried out their research with grants provided by the Council which did not exceed $10,000.

It should also be noted that there is a considerable concentration of scholarly interest in nineteenth-century science and technology: 35 projects, or 60 percent of the applications, focus on that period. The twentieth century has been studied in 20 percent of the cases.

The subjects of greatest interest in the history of science and technology have been biology, astronomy, engineering, and the philosophy of science. Next in line

are chemistry, technology, physics, and medicine, and then
various other disciplines. It appears that quite a wide
range of research interests characterizes the Canadian
community of historians of science. However, the highest
number of applicants in any one field is about five.

Now, as far as the *Canadian* component of the Council's
funding is concerned, I can report that, of 58 applications,
13 dealt with distinctly Canadian subjects. The remainder
dealt with mostly European topics followed by American ones,
with 4 percent of the rest defying geographical and political classification (e.g., philosophy of science). Statis-
tically, therefore, Canadian subjects constitute 22 percent
of all applications in the field. The total amount
awarded for Canadian subjects was $117,783 (which is 33
percent of the total amount awarded in the field), compared
to 40 percent for European subjects and 26 percent for
the rest. However, the success rate for Canadian studies
in history of science and technology was only 69 percent,
as compared to the general average of 84 percent.

According to our sample, proposals for studies concerning Canadian topics were received only from universities in Québec (Montréal, U.Q.A.M., Concordia), Ontario
(Toronto, York, Queen's), Saskatchewan (University of
Saskatchewan), and Alberta (University of Alberta), whereas
applications for the other subjects were received from
many other provinces, including Ontario, Québec, British
Columbia, Nova Scotia, Alberta, and Manitoba.

These figures, I must admit, contain some surprises.
The number of applications in the history of Canadian
science and technology over a period of eleven years amounts
to 13, submitted by 9 scholars. It is legitimate to ask,
consequently, whether this reflects the state of the history of science and technology in Canadian universities
and whether it is a sign of the way in which the Canadian
academic community expresses its view of the importance of
this field.

We are recognized as being an industrialized society,
and some are even predicting our membership in a post-
industrial society in the near future. One would imagine,

therefore, that Canadian scholars would have a particular
interest in the history of Canadian science. I imagine
that whether it is from a cultural perspective or from the
point of view of policy makers, a number of subjects would
deserve to be explored or to be examined more deeply. I
am not in a position to suggest what those areas of research or these priorities should be. But since it is one
of the objectives of this conference to discuss problems
of this nature to some extent, I am sure that you will
express your own wishes on the subject, and that you will
make known the problems you perceive and the needs which
you feel.

My statistics may have somewhat distorted the problem,
because universities may be awarding funds for research in
your area, funds which our Council is not being asked to
provide directly. Furthermore, there are no doubt various
other sources of research funds; provincial, private, and
other institutions, which are not included in my statistics.
I have not, for instance, taken into consideration research
projects carried out during sabbatical leave, or during
leave fellowship tenure. We cannot assume that statistics
taken from our research grants program are necessarily a
true reflection of the activities in a given field. After
all, I have not mentioned the dynamism which is evident
in your learned journals and conferences. We have not
undertaken any qualitative analysis of your achievements.
This would be presumptuous. Nevertheless, keeping in mind
the major role of the Social Science and Humanities Research
Council in the funding of research in the Humanities and
the Social Sciences, and referring to the results I have
pointed out previously, it may be worth asking ourselves if
there is room for improvement and for greater participation
on your part, or on the part of the Council. It may be
worth asking ourselves whether the universities and the
scientific community ought not to re-examine their role
and their commitment with respect to the needs and priorities of today.

This brings me to the second major point of my talk.
You are undoubtedly already preoccupied with this matter

and are presently undertaking a careful analysis of your
present activities and goals, with the objective of opening
up new avenues. You may have concluded, from what I have
said, that we suspect that there may be a state of stagna-
tion in research in the history of Canadian science and
technology. Should this be the case, I should like to
point out that the Council is interested in determining
whether it shares the responsibility for this.

In this connection, you will be interested to know
that the programs administered by the Council are about to
undergo an in-depth and systematic examination. Since his
appointment, the president of the Council has been assidu-
ously visiting and meeting with scholars, universities,
and learned societies in an effort to identify the needs
to which the new Council will have to respond. Our new
executive director is also very deeply committed to this
objective. I should mention that the staff of the Council
is being guided by the findings of the Healy Commission
Report, commissioned by the Canada Council in 1974.[2]

The program studies now being undertaken are a high
priority for the Council. They involve examination of the
impact of existing structures and activities on research
and Canadian society, an evaluation of their method of
operation, and an assessment of the results of the programs
in terms of our objectives.

The most appropriate thing for me to do in conclusion
is to ask you whether you feel that the Council has
measured up to your expectations up to now (keeping in
mind that the success rate is high and that the participation is overly low), and also to ask in what specific way
you feel we can adapt to better serve the goals which you
will establish for yourselves. In this context, I feel
I must point out that this community ought to be aware
that the Council, like other federal organizations, has
been subjected to budgetary restraints, and that the implementation of any new programs will need to be amply justi-
fied.

NOTES

1. Statistics provided by the Research and Analysis
   Section of the Social Science and Humanities Research
   Council.

2. Report of the Commission on Graduate Studies in the
   Humanities and Social Sciences, 1978.

ON PUBLISHING HISTORICAL STUDIES
Ian Montagnes

Mr. Mougeot has told us about the support of research, which is the beginning of publication. I shall skip over the next stage--the most difficult, most demanding, and most solitary: the act of writing. There has been a great deal published about this, but essentially it is something we must all face alone. I shall instead take up the story at the point where a manuscript physically exists. The problems it involves are those which face a Canadian scholarly publisher in making that manuscript readily available to the broadest possible readership--or in surviving as a publisher at all.

To begin with, some general observations: First, I shall be speaking of problems inherent in publishing any form of scholarship in Canada today. There is not yet a large enough sample to suggest that the history of Canadian science and technology has any publishing problems specific to it alone. In what follows, however, you may consider the probable effects of these circumstances: a discipline that is still forming; a core market for publications that is accordingly ill-defined, as compared, say, with economics where the professional membership in Canada is several thousand; a core market that is, moreover, defined by national interests--the history of *Canadian* science and technology--and thus largely by the boundaries of a small country; a more general potential market that is, by and large, illiterate in the subject of the discipline.

You will have noted the emphasis on marketing. This raises my second general observation, that even scholarly publishing is essentially a business. Dollars are neither our goal nor our raison d'être. Our rewards come from the excitement and the usefulness of diffusing new knowledge effectively and, when possible, elegantly. But our problems are largely financial, and it is the problems that are in the forefront this morning. Whether the

86

publisher is a commercial enterprise or a university press,
the problems are much the same. It is just as hard to
stay within a budgeted loss (that is, to stay within the
parameters determined by available subsidy) as it is to
achieve a budgeted level of profit. Only the x axis moves
up or down.

Nevertheless, there has by now developed a modest
corpus of works on Canadian science and technology. My
own press has published most recently *The Beginning of
the Long Dash: A History of Timekeeping in Canada* by
Malcolm Thomson, and before that, among other works, *The
Aquatic Explorers*, a history of the Fisheries Research
Board by Kenneth Johnstone; *The Miracle of the Empty Beds*,
a history of the conquest of tuberculosis by George
Wherrett; a history of biochemistry by Gordon Young; Ned
Franks's study of one specialized area of transportation
technology, *The Canoe and White Water*; the fascinating
wartime letters of C. J. Mackenzie and A. G. L. McNaughton,
edited by Mel Thistle; Mel's own early history of the
National Research Council (N.R.C.), *The Inner Ring*; the
histories of dentistry and of hospitals in Canada, by
D. W. Gullett and Harvey Agnew, respectively; and *A Heritage
of Light* by Loris Russell, who brought to the study of
technology the techniques of palaeontology and produced
what I think is one of the finest books we have published
at the University of Toronto Press. Both scientists and
technologists are also reasonably well represented in the
entries of the Dictionary of Canadian Biography/Dictionnaire
biographique du Canada, published jointly by Toronto and
Les Presses de l'Université Laval. Our interest in the
subject of this conference is indicated by the presence
of four members of our staff. Nor am I unmindful of books
published by other houses, in most cases commercially
oriented. Among these are J. J. Brown's works on Canadian
inventions; the documentary history of Canadian technology
edited by Bruce Sinclair, James Petersen, and Norman R. Ball;
the history of the NRC by Wilfrid Eggleston. There are a
number of biographies of Canadian doctors and scientists:
it is a comment on our time that the most popular figures

in print are those Canadian heroes of countervailing
ideologies, Alexander Graham Bell and Norman Bethune.

But to return to my topic, the problems of publication. The processes of publishing are described in this
booklet (*The Process of Publishing/ le livre universitaire:
avant et après*; Association of Canadian University Presses/
Association des presses universitaires canadiennes, 1977);
I shall not discuss them in detail. Of those most evident--
the copy editing of a manuscript, and its transformation
by printer and binder into a finished book--I shall say
nothing. They are important but are not central to publication. We can farm out the manufacturing process;
we can publish without either ink or paper, using film or
computer; we can even publish without copy editing. But
we cannot dispense with two other functions which are
absolutely central. The first is selection--what shall be
published. The second is distribution--ensuring that what
is published reaches the broadest possible readership that
can benefit from it.

In deciding what to publish, a university press does
not depend on its own editors' views, although they may be
the first test. We seek the best advice we can obtain,
from readers who may be in Canada or who may be half way
around the world, who are guaranteed anonymity so that
they will feel free to report frankly, and who donate time
and knowledge for the most modest of rewards. At its best--
and this is normal--the appraisal process provides many
benefits besides a simple screening. It relieves the
loneliness of writing to which I have already referred:
it is in effect a prepublication review. An editor may
make suggestions for revisions or improvement before a manuscript goes out for peer appraisal; the outside readers,
even when their reactions are generally favourable, almost
always offer suggestions for changes in organization,
emphasis, or detail. A good reader's report may run to
several single-spaced typewritten pages. The author need
not incorporate all the suggestions, which sometimes would
require another book, if not indeed another author. But
overall, as a result of this feedback, many works have

appeared that would not otherwise have merited publication, and many have appeared in a form greatly improved, to the benefit of both author and reader.

As to distribution, a publisher must be able to reach beyond the circle of the author's peers. He must maintain relations with wholesalers and booksellers who make the books available to libraries. He must be able to reach retail booksellers with books of general interest, and college bookstores with books related to courses. He must be able to carry books--including those that may be essentially national in interest--to an international market, and in these days of interdisciplinary studies he must be able to reach scholars in any field. He must have access to major bibliographical listings; know how to mount direct mail advertising campaigns effectively and economically; know when to buy display advertising; select the best places to send review copies; mount exhibits in Canada and abroad; be able to generate the word of mouth that is the most effective advertising of all. And he must have the backup staff to ensure that orders are filled promptly and accurately.

These two central functions are highly labour intensive. It takes time for a conscientious appraisal editor to consider a manuscript and pick the most useful experts to comment on what may be a highly specialized topic. And it takes time for an editor to work with an author in the development of a manuscript. It also takes time to plan and mount a sales and promotional campaign.

Time of course is money. Publishers need to be paid. And at money the crunch begins.

Like every other part of the world of higher education, the scholarly presses are facing austerity. To us, that is nothing new: we have been in this situation since 1972, when the prices of book manufacturing--for paper, ink, presswork, and binding--began rising at a rate far faster than that of the cost of living, at the same time as the market for books began shrinking. In the past six years, university and college libraries have reduced their buying for reasons that may or may not be financial. Unlike

the teaching wings of universities, however, the presses
have not been struck with a decline in the demand for
services. Student enrolments may be dropping, but new
areas of study--such as this one--keep rising, and the
submission of manuscripts continues. So does the pressure
from anxious authors caught in the web of publish or
perish. Thus the need for publication has been rising at
a time when we are less able to meet it, and when our
parent universities--who might or do assist publication--
are themselves hard pressed.

That Canadian scholars (in this new field and others)
do have a reasonable hope of publication--indeed, that
scholarly publishing in this country is as healthy as it
is--is partly due to the staffs of the various university
presses and the support received from university adminis-
trations. University presses and universities subsidize
the publication of scholarship to the extent of several
hundreds of thousands of dollars each year. It is also
in very large measure due to the Canada Council in years
past, and now to the Social Sciences and Humanities
Research Council of Canada, and to the two agencies they
have financed, the Humanities Federation of Canada and
the Social Science Federation of Canada. Grants from the
last-named body have made possible several works in the
history of science and technology. But there are dangers
in being dependent upon a Medici that is funded by Parliament, and no one can be complacent about the future of
budgets for research or publication.

Unless we can reverse the trends I have outlined--
rising costs, shrinking markets, rising pressures to publish, uncertain financial subsidies--then the publication
of scholarship in Canada will become increasingly difficult.
It will not die completely. There are books that will be
publishable no matter how difficult the situation becomes,
and it is a matter for great congratulations that the
number of university presses prepared to publish them is
still growing. But the marginal book--the book that is only
one more pedestrian increment to a well-studied subject,
the book that can expect to sell no more than 500 copies

over five years, the book in a discipline just beginning
to develop in Canada—these may fall by the wayside.

What can be done about all this? Let's leave aside
political action and the broad dynamics. Instead let me
suggest three things that individuals can do.

1. First, we can buy more new scholarly books. The
individual purchaser of a new hardcover scholarly book
appears to be an endangered species. Most people say this
is because university press books have become so expensive.
So, out of curiosity, a year ago I checked the cost of the
twenty-eight works published by the University of Toronto
Press in 1952, twenty-five years earlier. Their average
list price was $4.80. The average list price of the
twenty-eight new titles catalogued for publication in hardcover by the same press in the spring and summer of 1977
was $16.34. (But many of the titles were announced simul-
taneously in paperback at lower prices.) In this sample,
admittedly small, the price of a hardcover scholarly book—
probably the most difficult kind of book to publish in
Canada—had risen 3.4 times over a quarter century. This is
certainly greater, for reasons I have suggested, than the
rise in the cost of living over the same period. But it is
less than the increase in academic salaries at the University of Toronto during the same time. In other words, it
seems that academic staff have more disposable income than
they used to but are not applying it to their personal
libraries. If that situation were turned around—if, to
paraphrase something once said about poetry, every person
working on a monograph in history were to buy one new
hardcover monograph in history each year—scholarly publishing in Canada would be much healthier.

2. We can all write better books. I am not talking
only of subject matter although obviously a book of significance will always be more publishable than yet one more
dissertation in a well-ploughed field. I am talking about
the style of writing—the inconsistencies, fuzziness, non-
conforming usage in notes and bibliographies, infelicities,
errors in grammar and spelling, and questionable data that
manuscript editors now must ferret out at considerable

expense. At the University of Toronto Press, we calculate
our manuscript editing costs on the basis of salary,
fringe benefits, space, light, heat, and other overheads
which cannot be ignored, even if many of them at other
presses are hidden in the budget of the parent university.
It comes to about $12 per hour. A manuscript may take
up to 100 hours of editing. That time often could be
reduced if the author had prepared his manuscript properly.
The charge for the extra time should in fairness be levied
on the author; but it now is being passed on to the Press's
own subsidizing fund, or to a subsidizing agency or the
purchaser.

　　　3. We must be prepared to accept alternative forms
of publication for those works that can no longer be published in the traditional manner within acceptable levels
of subsidization. I am not speaking of a proposal for
increased publication through learned societies, for that
would still employ the traditional methods of publication--
and differs from what I have described only in that the
editorial function is passed on to the society and the
distribution function is restricted. Rather, I am suggesting that we are all going to have to get used to microform publication of new works in our fields--particularly
publication by microfiche, which is much more convenient
than the old roll microfilm, and much less annoying to
use. We may find more and more publication by electronic
means as well. We shall be hearing more about on-demand
publication, in which copies literally are produced, in
microform or xerographically, to meet orders as they come
in. One may predict that the journal, as a form of publication, will change dramatically in the next two decades.
But I am most taken by a proposal published in the last two
issues of *Scholarly Publishing*[1] which would extend a service we are all familiar with--the storage and distribution
of theses through the National Library in Canada and
University Microfilms in the United States. The suggestion
is that this be applied to new works as well. New works
would go through an appraisal process but instead of being
published would be deposited in one or more central

agencies. The agencies would advertise the availability
of these works through subject catalogues and other means:
bibliographical control would be essential to success.
Scholars wanting to read the works could then obtain them
at relatively low expense, either in microfiche or in bound
volumes printed xerographically on demand. The central
agency would in short function more like a library than a
publishing house; and it would ensure the availability
of works that could not be published traditionally.
There is a great deal more to be said about alternative
means of publication, but this relates more to future
answers than to current problems, and should perhaps
be discussed in the question period or on another occasion.

NOTE

1. August Frugé, "Beyond Publishing: A System of
   Scholarly Writing and Reading," *Scholarly Publishing*
   9:4 (July 1978), 291-311, and 10:1 (October 1978),
   17-35.

SCIENCE AND A CULTURAL FUTURE:
THE USEFULNESS OF THE HISTORY
OF SCIENCE AND TECHNOLOGY
TO DECISION-MAKERS
H. R. *Wynne-Edwards*

The Age of Reason

We are unable to define the age we are entering or
even the one we live in without the perspective of history
from which to review it. Could it be the Post-Industrial
Age foreseen by David Bell and others--one in which services rather than goods become the dominant economic
traffic? Might it be a further Nuclear-Industrial Age of
great technological wonder--the threshold of the world of
Star Wars and Battlestar Galactica which have so captured
public imagination? Could it be the Dark Age that many
have predicted--a time when the economic system of the
western world has collapsed, when the lights have gone
out, the public institutions are shuttered, and man has
once again to wrestle on a personal basis with his needs
for shelter and a food supply?

Personally, I think it will be none of these, but an
age in which man is confronted with his physical limits
of existence. It will be an Age of Reconciliation between
natural law and human culture. If Star Wars becomes a
reality it will be because a selected few have left this
planet behind, abandoned as inadequate for more than a
modest launching base. Even as they carried our illusory
conquest of nature into outer space they would be confronted, not necessarily by their own limitations, but by the
limitations of the life-supporting system of nature itself.

If the Dark Age becomes a reality it will be because
we shall have failed in our act of reconciliation, frozen
and immobilized in the path of our economic momentum in a
society so complex that neither the political nor the institutional will can be found to initiate the necessary

94

changes. Star Wars can be taken as one extreme limit of
the possible future, a simple extrapolation of the
"business-as-usual" world of the technological fix. The
Dark Age can be taken as the other extreme--the ultimate
"back to nature" scenario of the most ardent conserva-
tionists, under which perhaps ten percent of the present
world population could find some primitive sustenance,
having, perhaps, deployed the balance of society as or-
ganic fertilizer.

There can be some confidence that the real future will
lie between these extremes. There are enough signs already
that we can adapt quickly to new ideas and new values. I
see no cause to despair.

Thus whatever it is, the future will necessarily also
be an Age of Reason. We are already beginning, not just
to examine the course of progress, but the very idea of progress itself. We are in what might be summed up as an
ecological predicament--one in which the scale of human
activity competes with the scale of the natural cycles which
renew the stocks of minerals, fresh air, energy, and pure
water. As surely as we extend our technologies to enhance
the supply of one, as surely we shall jeopardize another.
*The more we act upon our environment, the more our environ-
ment acts upon us.*

We have known this for many centuries, and as noted,
it was probably much clearer at times to our predecessors
than to us. The science in the wake of Francis Bacon
offered a promise of mastery over nature and the Industrial
Revolution appeared to be that promise fulfilled. For two
hundred years we have pursued that illusory pot of gold.

A series of "crises" (as we have come to call them)
have had the effect, in the last decades, of bringing this
physical reality to public consciousness. In succession,
since the early 1960s, we have started to become birth-
conscious, and ocean-conscious. A few more extreme winters
and summers, misplaced monsoons, and parched reservoirs
will serve to make us climate-conscious. Each of these
crises, triggered more often by man than by nature, has
been like the scattered tips of some huge floating iceberg--

signals of something much more profound underneath. The
iceberg itself needs to be dragged ashore and revealed to
view. That will finally make us Earth-conscious. Earth-
consciousness will force us to acknowledge what we have
known since Columbus: that we inhabit the delicate, almost
two-dimensional surface of a small and spherical planet.
It is all that there is. It is the only known place in the
universe where sunlight, air, water, and soil meet and in-
teract in that particular and fragile balance that supports
the only biosphere we know. Being Earth-conscious will
hopefully give rise to a new morality: a set of physical
ethics to parallel the social ethics that regulate man to
man.

This is a very large subject which I shall only touch
on here. It will perhaps suffice to say that the Faith,
Hope, Love, and Charity which serve as guiding social concepts may soon be joined by Endowment, Integration, Balance,
and Obligation as the conceptual basis to regulate man to
Earth.

Already, however, we can see that these renewed and
emerging human values now expressed in conservation, environmental protection, and resource management are as yet
fragile luxuries willingly and quickly sacrificed on the
altar of economic necessity. The present economic-energy
recessions are teaching us that these values exist only
because of a thriving industrial base that provides the
wealth for us to indulge in them. Jobs are more urgent
than pollution controls, nuclear power is more urgent than
the problems of radioactive waste management. Economic
growth is more urgent than conservation. In the intricacy
of modern human affairs, the urgent replaces the vital, the
short-term need outweighs the long-term gain. But at each
juncture when such sacrifices are made, the crisis intensi-
fies and the satisfactory solutions escalate in cost and
complexity merely by being postponed.

We know that there are as many jobs in conservation,
efficiency, and re-use as there are in primary production;
that services can substitute for goods in the equation of
individual well-being, and that dynamic change and growth

would be fully as possible within a finite envelope as within an expanding one. A mature forest does not continuously encroach upon the skyline, although there is continuous change within its internal economic system. Intellectually we can envisage a world in balance with its endowment and in harmony with its physical system. It would be one of fixed but rotating stocks in which the only continuously expanding quantity--the last exponential function--was that of knowledge itself.

Having conceptualized this kind of world, how do we make the necessary transition? Possessing the knowledge, how do we apply it? Do we learn by looking back, or by looking ahead? How, in short, do we get from here to there?

The Age of Economic Growth

M. King Hubbert, the geologist who forecast the energy crunch with uncanny accuracy in 1956, has gone on, as have many others, to study the consequences of forms of cumulative production other than that of oil and gas. Until a century after the Industrial Revolution began, economic growth was a slow and uncertain quantity, pegged to the land in cultivation, and expanding only as trade, and therefore effective access to that land, itself expanded. The Industrial Revolution had many roots, and has been attributed to many causes, including some that were much more probably effects, like capitalism. To the reflective scientist, its root is almost certainly the steam engine, which provided the first significant departure from human dependence on solar energy in the form of sweat and sinew.

It is easy to build spurious correlations from the similarity of various hyperexponential growth curves from then on, but the curve of man's access to power or energy, his rate of consumption of materials, the rate of growth of capital and interest, and above all, the expansion of the human population, are parallel functions. They establish a direct link between energy technology and economics that the current recession reveals is still very much with us.

From a linear correlation between energy use and
GNP prior to 1973, when energy cost was a negligible and
thus ignored input to the economic cycle, the pattern ex-
ploded with the quadrupling of oil prices. The United
States and Canada, the oil junkies of the modern world,
have since fallen from first and second to seventh and
ninth places in terms of per capita wealth of nations,
being surpassed by not just the oil-rich Arab States, but
by the more energy-efficient economies of western Europe.
Energy in the quantities demanded will never be cheap again.
Economic growth in the form we knew it will not be in our
reach again. We would do well to do our accounting of
human activity in energy transfers as well as cash flows,
and our measurement of social efficiency less with cost-
benefit analysis than with the laws of thermodynamics.

We have surely relied too long on economics as the
science of human traffic. Economics since Adam Smith has
defined human activity as a manmade, closed cycle governed
by labour and capital, and virtually independent of the
seemingly infinite stocks and sinks represented by the
physical world, which were there to be called forth as
demanded. For this reason most of us, even experts, have
yet to make the connection between the energy crisis of
1973, inflation, and the relative performances of curren-
cies. If the concurrence of inflation and unemployment
confounds conventional economics, it is because conventional economics has overlooked yet another "unseen hand," that
of science and technology. It is the hand that has pushed
us from cave to condominium. If it got us from there to
here, it can equally surely get us from here to wherever
it is we want to go.

Conventional economics reveals that market forces dictate the pattern of industry, demand calling forth the
necessary scale of production and supply. How is that
demand established? We could extend the expensive energy
lesson we are presently being taught to postulate that it
is in fact a technological compulsion that has propelled
economic growth and development. If steam power induced

the other factors of production, did not the transistor more recently propel across world markets the digital watch, the pocket calculator, the microprocessor, the ubiquitous radio, and the electronic bug? Do you recall "demanding" the supersonic transport, the neutron bomb, or the feminine hygiene spray? How on Earth, I say, can the technological push behind the so-called market pull have escaped our general attention?

Culture

Our culture is an aggregate of what we are, what we value, what we think and what we do. For perhaps 6000 years we have had ancestors who conducted themselves in ways quite recognizable today. *Science is our only distinctive culture.* The largest differences in the medium of art, the sound of music, and the forms of communion and communication are those engendered by our technology.

One would never guess that this were so from current affairs. Turn the pages of a weekend newspaper. Where is the news of science that makes us what we are? What colossal hoax is this, and how and by whom is it perpetrated?

Those responsible for the media would explain that market forces operate and that the public get what they demand. They are not, one would therefore conclude, demanding science. Yet a bemused CBC report lists "The Nature of Things" (one of only two science programs) among its thirteen most popular regular features, with listening audiences of over one million. It has roughly the same audience level as the National News, and half the audience of Hockey Night in Canada. Science Museums like the Ontario Science Centre in Toronto and the Museum of Natural Science in Ottawa, I am told, attract three to five times the visitors of neighbouring art galleries, and a vastly greater audience than do ballets, drama festivals, and symphonies to which large public subsidies are applied. I would conclude that there is yet another unseen hand here that shapes the pattern of cultural demand in our society-- the cultural push of traditional forms of entertainment.

J. Tuzo Wilson has pointed out that science alone in

the catalogue of cultural activities does not require an audience to survive as do, for example, art, music, drama, ballet, and religion. Because its ultimate product is technology, there is no apparent need separately to advertise or market its cultural value. Hence the number of people engaged in "pushing" science through advertising and entertainment is negligible. Those few in society that understand science have other forms of reward and hence are otherwise engaged. The cultural world of communications is thus vacated in favour of those who do not understand or acknowledge one of the main cultural forces that has shaped our society. We drift sideways through a fog of obscurity on an unseen technological current.

I lay most of the blame at the door of education and the way in particular that science has been introduced and taught to children. Again it is a long story which I shall only summarize, but I am prepared to offer a protracted and spirited defence of its foundations. In essence, it is that a child, left to itself, would first experience its physical world in the form of creature comforts--food, warmth, and shelter. These would be largely artifacts in a manmade envelope, now institutionalized in the applied sciences of agriculture, medicine, engineering, architecture, and the like. Next it would encounter the world of nature--clouds, rocks, soil, water, trees, plants, and animals. In terms of education, the child is scaling a pyramid of science, from the applied sciences at its base to the biological and earth sciences. Soon would come reocgnition of some relationships and essences: mass, form, symmetry, number, class, order, action and reaction. It has ascended higher to enter the domain of physics and chemistry. Finally would come the unifying and simplifying abstractions of geometry, algebra, and arithmetic. This is mathematics, the apex of the pyramid of experiential learning. The child has reached it naturally by successive steps of encounter, discovery, and rationalization.

Of course this pyramid for the purposes of new discovery is equally a pyramid of dependence, in which mathematics alone among the physical and natural sciences can progress in absolute abstraction. Physics and chemistry depend on the fundamentals of mathematics, the earth and life sciences on the fundamentals of physics and chemistry, and the applied sciences, in turn, on the whole pyramid above. From this realization we have come to define the apex of the pyramid as the basis of science, and the base of the pyramid as its derived and dependent end. We have thus inverted the natural pyramid of knowledge, simultaneously organizing the system of reward for scholarship (apart from medicine, which is too close to home) in the same order.

Because of this sophisticated and intellectually elegant inversion we have chosen to drive the pyramid of science into the heads of school children point first, with a stiff preliminary dose of "basic" mathematics and sciences taught in isolation. The result is so alienating to most children that only a very few persist to become scientists or engineers (in which case they just might rediscover the "real" world around them at last). The rest, the vast majority, subconsciously eschew the entire process, and being unable to make the necessary connection, discard their abstracted scientific education as unrelated or not "relevant" to the world in which they must find a place. They then go on to become our businessmen, our managers, our lawyers, our communicators, our politicians, and also (alas!) our economists. The result is the predictable one I have been describing. Instead, we could attempt to provide a common basis of experiential understanding of the physical world in primary and secondary education, from which the inevitability of an approach to scientific specialization through basic mathematics, physics, and chemistry at the tertiary level could be more readily understood and accepted.

The History of Science and Technology

To be perfectly blunt, I don't think that the history of science and technology has more than the most marginal impact on the current course of public decisions. I also think that that is one of the reasons why the decision-making, planning process seems so inadequate to our times.

The history of science and technology should be of paramount importance. It is not. Yet it is central to our understanding how we got to be where we are today.

The Age of Reason will have to be an age of science. Planning, forecasting, analyzing, synthesizing, and rationalizing have become central activities of government and governance is becoming the central activity of society itself. If that governance does not include the medium and long-term perspective afforded by scientific knowledge, and if that knowledge does not include the perspective of its own historical evolution and the social compulsions in its wake, I do not think we will ever get "from here to there," even when "there" is the only way to go that leaves the human option open.

An improved grasp of the role science and technology have played in human culture will help us manage our accumulating knowledge more directly in our own interest. To some extent we have been at the mercy of our technology and to some extent its servant. Science has no doubt caused problems as well as solved them.

If we can devise the proper means (and because of public pressure it is already evolving) we can choose among the options presented by the inevitable expansion of knowledge and its resulting technology, setting aside what we do not want, and selecting only what is beneficial. The means thus involve an evaluation not just of our capability of doing something but also the desirability of doing it in social and environmental as well as economic terms. Mere capability should no longer inevitably dictate implementation--there are many things we could but should not do.

That these decisions are difficult does not mean that they should be avoided. Genetic engineering, nuclear power, northern development, and supersonic transport are current examples of issues that have become sharply polarized, because the means of addressing them and achieving consensus are not yet in place.

Getting there can be half the fun. If the technological push represented by man as discoverer, doer, and maker has in fact propelled us inexorably from the nomad hunter

and gatherer to the urbanized world citizen of today,
and if market pull is in fact dictated by the same technological push, then all we have to ensure is that it is
a push in the direction we want collectively to see taken.
The new-found unseen hand, once recognized for what it is,
can project us in any direction we choose because this
hand at our backs is our very own.

Science and technology can be directed at social as
well as economic or military ends. We can be ejected into
space for a brief time to make the Star Wars dream reality,
we can be pushed over the edge of no return to a Dark Age
on an exhausted and corrupted planet, or we can pick a
middle road that harmonizes our aspirations with our
ecology. So the question is not really how to get there
from here because we possess the means of any ends. The
hardest part will be persuading each other that it is time
to start and that the ends will be worthwhile. Deciding
where we want to go may be the hardest part of all.

The profession of the historian of science is a
bridge between the two cultures of humanities and science.
It could do much to help these cultures understand each
other and eventually coalesce. It could help us see the
unseen hand, and having seen, move on.

IV

EPILOGUE

AFTER-DINNER THOUGHTS
ON THE INFRASTRUCTURE OF SCIENCE
J. W. T. Spinks

In 1962, a book by T. S. Kuhn appeared entitled *The
Structure of Scientific Revolutions*. Among other things
it drew attention to the change in approach of philosophers
and historians of science who now favour perceptions of the
scientific enterprise that take human factors into account
as well as the purely logical structure. According to
Kuhn, science consists of a series of peaceful interludes,
punctuated by a set of theories, standards, and methods
which he refers to as a "paradigm" (Webster; *paradigm*--
"pattern, an example"). Actually, what Kuhn is doing is
exploring the process of arriving at a theory. He is
constructing a *sociological* theory of scientific theory
construction. In modern parlance he is describing the
infrastructure to one part of science, the *theory* construction part.

Obviously, there are many parts of science--doing
experimental work, making observations, putting forward
tentative hypotheses, testing these hypotheses, applying
scientific knowledge to everyday life, etc., etc., and
each of these parts will have an infrastructure.

The actual infrastructure which evolves will depend
on the goal to be achieved but whatever the goal, assuming
that there is a goal, reaching the goal will certainly be
helped if one knows something about the matter in hand.
This begins to sound suspiciously like a systems approach
and of course, that is exactly what it is. But don't
worry, my name is not Glassco or Lamontagne and I am at
heart a pure research scientist and have always spent a
good deal of time sniffing out the most academic approach
possible to whatever I happened to be doing. I remember
getting a student to study autochromatography, the chroma-
tography of a substance by itself, in which he poured saturated $BaSO_4$ solution through a column of solid $BaSO_4$ in a

chromatographic column--of course saturated $BaSO_4$ solution
came out at the bottom. The problem seemed at first sight
to be completely pointless but it took on a good deal more
meaning when the $BaSO_4$ solution was tagged with radioactive barium and sulfur which then enabled us to study
the surface exchange occurring. Incidentally, it leads,
some thirty years later, to some very pertinent thoughts
about the removal of radium contamination from runoffs
from uranium tailings.

The building up of a more general infrastructure to
pure and applied science as a whole is a worldwide phenomenon. In Canada, this growth has occurred gradually in the
last fifty or sixty years and is a matter of considerable
national interest and importance. I had the good fortune
to be involved in the growth of some of the elements of
this infrastructure. Some would say that I gradually fitted
in with the bureaucratic process and became a science
bureaucrat. At all events, I propose to spend a few minutes
telling you some of my impressions of the present infrastructure, as perceived over a period of some fifty years,
beginning with the research process itself.

My first serious brush with scientific research came
exactly fifty years ago. In the summer of 1928 I received
a postcard from Paris, from Professor A. J. Allmand, a well-
known photochemist. It was brief and to the point. "My
dear Spinks, In a few days you will hear that you have
been awarded your degree--Congratulations! As far as I
know you have not yet made up your mind what you are going
to do. How about working with me? If the answer is yes
you will find that there is a small lab set aside for you
not far from the College Machine shop. See you in the
Fall. P.S. I think I can scrape up enough money for you
to live on." The matter was discussed with my mother who
said that I could do what I liked, provided I saw after
myself financially. So a few days later I packed my bags
and went to King's College, London, with the general idea
of spending a year or two learning what it was to do research.

The problem that I was to work on was indicated on a

card, lying on a table in my lab. The Professor had evidently felt quite sure that I would fall for the idea of working with him. Anyhow, the card said, "Look up a paper by Weigert in Annalen Physik 1907, p. 243 and try to reproduce the experiments. It might be worth reinvestigating!" The paper described a photochemical experiment in which ozone was mixed with chlorine gas and illuminated with blue light.

I set to making equipment in which the experiment could be carried out at various temperatures and gas concentrations and with various wavelengths of light. I learned to use a lathe and do simple machine-shop work, to blow glass and standardize light intensities, and eventually, after four or five months I started to obtain some crude results. During all this time, not a sign of my Professor except that he had arranged for me to receive a grant from DSIR for about 150 pounds a year. Also at his suggestion I sat examinations for a London open scholarship worth 100 pounds and a London Exhibition worth 80 pounds a year for two years. I was fortunate enough to win both and jumped from rags to riches almost overnight.

As soon as I started to get results the Professor started to drop in once a day and look over my shoulder at the graphs I was drawing. The first time, he withdrew his pipe slightly, emitted a "huh," and walked away quickly. However, after another month or so, the results started to become reproducible and one could begin to place some confidence in them and the Professor would sit down to chat. Eventually two rather substantial papers appeared on the mechanism of this particular photochemical reaction.

My two years of research with Professor A. J. Allmand were most rewarding. He was one of the leading photochemists at that time, which meant that he not only attracted good people to work with him but a number of visiting firemen were constantly dropping in. We students thus had the benefit of discussing the latest work going on in various labs at first hand with the actual people either doing or directing the work. We did not attend any formal classes but there were enough research people around that

one was constantly stimulated to dig quite deeply into
other areas of knowledge. As an example of this sort of
thing I met a physics student named Johnson who was doing
spectroscopic research in a lab not far from mine. He
invited me to see his equipment and told me quite a bit
about the whole subject of spectroscopy.

Then, one day in the summer of 1930, Professor Allmand
came to my lab and told me that a friend of his, Dr.
Thorvaldson by name, was coming over to Europe from Canada
to visit Iceland, his birthplace, and that he was looking
for a junior person to work in his chemistry department in
Saskatoon--at the University of Saskatchewan. Would I
like to meet him? Yes, I would and a meeting was arranged.
A few days later Dr. Thorvaldson told me that Dr. Murray,
the President of the University of Saskatchewan, had authorized him to offer me a position at the University. By
this time I had complete faith in Dr. Thorvaldson and I
said yes--having really little or no idea what Canada or
Saskatoon or the University were like.

I was told that I should plan to reach Saskatoon by
mid-September and this meant doing everything in quite
a rush. First I went home to consult with my mother. She
came to the point right away--"are you asking me or telling
me?" I said that really I had made up my mind and was
telling her but that I hoped she would agree--and she did.

My Professor then came into the picture. Since I
was going to Canada, he thought that I had better make myself legitimate and get my Ph.D. Had I been staying in
England, a Mr. would have been quite adequate. He then
told me that he had taken the precaution of registering
me for a Ph.D. degree some time before. If I could write
up a report and get him five or six copies he would arrange
with the authorities for an oral examination. Luckily
by this time I had written a number of reports about my
work and had no trouble in putting together a suitable
thesis--adequate but not very elegant would probably best
describe it.

On the appointed day I turned up at my Professor's
office and met the examining committee--Professor Cyril

Hinshelwood from Oxford, Dr. Norrish from Cambridge, and
Professor Heilbron from Liverpool, all seated round a
small table with my Professor. It was a pretty formidable
examining committee as you will gather from the fact that
both Hinshelwood and Norrish were eventual Nobel prize
winners in chemistry! My Professor started by suggesting
that Professor Hinshelwood might like to ask me a few
questions about my thesis. "Nothing doing," said
Hinshelwood, "it's quite clear that Spinks knows far more
about this subject than any of us here, including you. Why
should we make ourselves look silly? But don't worry, I
do have a few questions to ask." And he proceeded to ask
questions in considerable depth about matters related to
the topic. I remember him asking quite innocently what
the physicists meant by P, Q, and R branches in a spectrum.
Luckily I had discussed this sort of thing with my spectroscopic friend Johnson and could give quite a good answer
and he was duly impressed since in 1930 this had not yet
got into the textbooks. Eventually he asked a question to
which I had to say "I'm afraid I don't know"! "Well,
neither do I, so we are even and you are honest--that's
all I want to ask."

Norrish and Heilbron went through the same routine,
each taking me to the point where I had to admit that I
didn't know. At this point, about two hours after the start
of the meeting, my examiners exchanged glances and in turn
tapped the table. Whereupon my Professor leaned forward,
picked up a small silver bell and rang it. His Secretary
poked her head around the corner of the door--"yes,
Professor?" "Tea please, Alice," and that is how I got
my Ph.D. Actually, it was a foregone conclusion, as I
have discovered since that my Professor didn't recommend
anyone for the oral until he felt absolutely sure that
they would pass with flying colours.

A few days later I was on the Empress of Australia, on
the way to Canada and the University of Saskatchewan. I
was twenty-two years old!

I have told this story in some detail since it illustrates the human aspect to two papers which eventually

appeared, entitled "The Photosensitized Decomposition of Ozone by Chlorine," and which completely neglected the human side of the story and said nothing about one of the essential ingredients of science, the creative process.

At the University of Saskatchewan I gradually got a program of research going, at first in photochemistry, but later in spectroscopy with Herzberg in Darmstadt, Germany, and still later in operations research and atomic energy. Each piece of work had a very personal side to it as well as what we might call the scientific side. And I became involved with students, grants, meetings, journals, fellow scientists, and so on.

A quickie--related to C. J. Mackenzie. At one stage of my work with the Canadian Atomic Energy Project we wanted to put up a pilot plant for recovering plutonium from irradiated uranium slugs. Everything had to be done by remote control and it appeared that the cost might amount to some hundreds of thousands of dollars--quite a bit for those days. The director of the Chemistry Lab, Dr. Steacie, suggested that I take the matter to C. J. for his approval, so a meeting was arranged. Dr. Mackenzie immediately started talking about Saskatoon, the University, the war, etc., until I began to get worried and said that I knew that he was very busy, would he like to glance at the proposals. He took out his pen and quickly signed the document. I was horrified and said, "But Dr. Mackenzie, don't you want to hear something about?" "Shall I cross out my signature?" asked C. J. "Well, no." "OK. I suspect that you and Dr. Steacie are prepared to back up the document and that is good enough for me. But, if it turns out that you have made a mistake don't come back to me again." Dr. Mackenzie believed in delegation of authority and I applied the technique on numerous occasions after I became president of the University. Gradually, too, I became involved in the organizational aspects of science. Probably my first introduction to science bureaucracy was as a member of a University Council Committee on Graduate Studies. In 1946 the Committee was transformed into a Faculty with Dr. Thorvaldson as Dean. I succeeded

Dr. Thorvaldson as Dean and one of my first problems was
to expand graduate work from Master's to Ph.D. level. I
quickly realized that the level of research activity in the
University was almost directly proportional to the funds
available, so when I was asked to serve on research bodies
I usually said yes. Each time I learned quite a bit and
gradually acquired a feel for the organization of research,
particularly at the university level. Organizations and
commissions with which I was connected in the 1945-75
period include the National Research Council, Saskatchewan
Research Council, Defence Research Board, Canada Council,
Oil & Gas Conservation Board, Centre for Community Studies,
P. A. Newstart, P.O.S. Pilot Plant, SED Systems Ltd.,
Institute for Northern Studies, Institute of Pedology,
PICUR, AUCC Board, Banff School of Advanced Management,
Canada West Foundation, etc. etc. and so forth. Each
time, matters of organization and financial support were
matters of great consequence even for those doing the
purest of pure research or engaged in writing the most ab-
stract and scholarly book. I remember, for example, while
I was a member of the National Research Council, I was involved in a report on "Forecast of Needed Federal Support
of Research in the Natural Sciences and Engineering in
Canadian Universities, 1964-69" and in getting major equipment grants of the one-million-dollar order. At the
University of Saskatchewan we were interested in getting
a linear accelerator costing about one million with a
building, also about a million. I broached the matter
with Dr. Steacie and at an appropriate time Dr. Steacie
introduced the matter at a meeting of the Council of the
National Research Council. Members of the Council were
interested but worried lest the money would come at the
expense of ongoing programs. Dr. Steacie said that nothing
would be done until extra treasury support was guaranteed.
Eventually Dr. Steacie announced that approval had been
obtained for extra funds and he added that curiously enough
Spinks had just put in an application for three-quarter
million. The other members of Council were momentarily
taken aback but before they could get mad, Dr. Steacie

added that of course nothing would be done until all
universities had had a chance to apply and that all
requests would be put through a committee on which the University of Saskatchewan would not be represented, Ha Ha Ha,
and the whole thing went off pleasantly. Actually, in
the outcome, the University of Saskatchewan got its three-
quarter million but subsequent requests amounted to much
much more so no one lost out from our trail-breaking
exercise. The whole thing emphasized the foremost importance of good leadership--in this case from Dr. Steacie.

There is room even in erudite histories for anecdotes
of this kind which don't appear in Council or Board minutes
and certainly not in scientific papers!

Now it's time to stop. If time had allowed I would
have expanded the infrastructure theme, arriving at a
ladder or chainlike structure linking the creativity of
the scientist to the social changes eventually resulting
from the discoveries of the scientist or, in reverse, going
from desired social change to the creativity of the scientist. It would look something like this:

New ideas----------------------creativity
Financing of research-----------facilities, funds
Systems approach----------------information retrieval,
                                research inventories, re-
                                search programs, research
                                groups, feedback, societies,
                                joint meetings such as this
                                one, etc.

Strategy to obtain sub-goals----Research Councils (national
                                and provincial), Industry,
                                Universities, etc.

Sub-goals-----------------------Provincial Science Councils
Overall goals-------------------Science Council
Overall policy------------------Government (Ministry)
Social Change

In using the chains we need to pay attention to the
links--including some I have left out. This infrastructure
will vary from time to time, from place to place, and from
discipline to discipline. From the history of science point
of view, it provides a rough framework of reference against

which happenings in science can be evaluated, the reference frame being part of a systems approach.

REFERENCES

1. T. S. Kuhn, *The Structure of Scientific Revolutions* (Chicago: University of Chicago Press, 1962).

2. T. S. Kuhn, *The Essential Tensions: Selected Studies in Tradition and Change* (Chicago: University of Chicago Press, 1977).

3. J. W. T. Spinks, "Gift of the Gods," *Chemistry in Canada* 27:5 (1975), 21.

4. J. W. T. Spinks, "Science and Social Change," *Chelsea Journal* 4:1 (1978), 30.

APPENDIX A
APPENDICE A

PROBLEMS AND ADVANCES IN THE HISTORY OF
CANADIAN SCIENCE AND TECHNOLOGY

PROBLEMES ET PROGRES DE L'HISTOIRE DES
SCIENCES ET DE LA TECHNOLOGIE CANADIENNES

WORKSHOP REPORTS/
RAPPORTS DES ATELIERS

INTRODUCTION

The eight workshop papers and their commentaries constitute a report on the state-of-the-art in the history of
Canadian science and technology. That they cover a wide
range of interests is a reflection of the multifaceted
nature of this enterprise. It encompasses the intellectual
side of history, as represented by the reports on discovery
and invention, adaptation and innovation, and institutional
frameworks. On the other hand, there are pedagogical
issues for the new discipline, represented by the workshops on teaching materials and course structure and content. Two deal with important ancillary activities, those
of archives and museums. Finally, the workshop on the
history of medicine touches on aspects of all of these
problem areas.

While the reader may not immediately see much coherence in this set of reports, there is a theme throughout:
the field of the history of Canadian science and technology
is not a unitary field, such as Canadian history, in which
there may be considerable diversity but a common stock of
problems and identifiable approaches. This field is too
young; therefore, work goes on along many fronts and no
synthesis has been effected. Indeed, it may be many years
before anyone would attempt such a synthesis. In lieu of
one, the reader must have some patience, but consider the
problems that confront each group as it defines its part
of the whole. The problems which all the groups share and
their common enthusiasm, must constitute the coherence of
the field for now.

The papers have been edited only slightly; the commentaries have, in most cases, been shortened somewhat. The
discussions, as reported to the plenary session by the
rapporteurs, have been paraphrased and commented upon by
the editors.

INTRODUCTION

Les communications et les commentaires qui ont été
lus dans le contexte de huit ateliers différents consti-
tuent un bilan de ce qui se fait de mieux en histoire des
sciences et des techniques canadiennes. Les textes qui en
résultent abordent des questions très diverses, ce en quoi
ils traduisent les nombreuses dimensions de ce projet
global. En effet, il comprend l'histoire envisagée comme
histoire des idées: découvertes et inventions, processus
d'adaptation et d'innovation, cadres institutionnels. Par
ailleurs, cette nouvelle spécialité fait face à des problèmes d'ordre pédagogique et dont les enjeux se trouvent
au niveau des matériaux didactiques, de la structure et
du contenu des cours. Deux ateliers se sont occupés des
sciences auxiliaires de l'histoire, à savoir des archives
et les musées. Enfin, l'atelier portant sur l'histoire de
la médecine recoupe tous ces genres de problèmes.

Le lecteur ne percevra peut-être pas immédiatement en
quoi consiste la cohérence de ces différents ateliers, mais
il existe pourtant un thème commun qui leur est sous-
jacent: le domaine intellectuel que constitue l'histoire
des sciences et des techniques canadiennes n'est pas struc-
turé comme l'est l'histoire du Canada où, en dépit d'une
variété considérable de styles, des problèmes communs et
des façons communes d'aborder les questions existent.
L'histoire des sciences et des techniques canadiennes est
trop jeune pour bénéficier d'une telle armature intellec-
tuelle. Dans l'attente de grands travaux de synthèse,
les chercheurs occupent des créneaux divers et non connec-
tés entre eux. Il faudra peut-être attendre des années
avant qu'une synthèse ne puisse être tentée en ce domaine.
Le lecteur devra donc se faire une raison face à cette
lacune et se montrer patient. Il faut d'ailleurs bien
mesurer les problèmes auxquels chaque groupe de chercheurs
doit faire face tandis qu'il tente de définir son lieu au
sein du projet total. C'est dans ce sens que les groupes

se ressemblent, en plus de partager un enthousiasme commun, et c'est ce qui doit servir de fil conducteur pour explorer l'histoire des sciences et des techniques canadiennes dans l'état actuel des choses.

Les communications n'ont subi que de légères retouches. Dans la plupart des cas, les commentaires ont été quelque peu raccourcis. Quant aux débats tels que rapportés à la séance plénière, ils apparaissent ici sous forme résumée, accompagnés d'ailleurs de commentaires insérés par les responsables de ce volume.

THE ROLE OF THE INVENTOR IN CANADA AND
THE USE OF PATENT OFFICE RECORDS

Loris S. Russell

Introduction

The terms invention and discovery are commonly linked
in historical studies of science and technology, discovery
being thought of as the basic process and invention as the
practical application. But by definition the two operations are quite distinct. Discovery is the disclosure or
revealing of some thing or principle that already exists,
whereas invention is the finding out or contriving of something new, something that previously did not exist. On
this basis invention might be looked upon as a more fundamental process than discovery, even though invention is
based on knowledge of observation already acquired.

The great inventions of prehistoric times were probably made at more than one time and place. Early man observed the phenomenon of fire and found that he could use
it to warm himself and to cook his food. These were discoveries. When he devised methods of creating fire, either
by intense friction or by percussion of objects against
each other, he had made an invention. The fact that sharp
stones could be used as cutting and piercing implements
was a discovery, but the technique of chipping these into
an artificial cutting edge was an invention. The wheel was
probably the first mechanical invention. We can imagine
that prehistoric man observed that logs rolled downhill,
and that by placing them under a heavy object, such as a
large boulder, he could use this rolling motion to help
him move the object. The real invention came when he devised the axle and the bearing, so that the roller could be
carried along with the load.

Early in historic times man found that he could heat
certain minerals and get metals that could be shaped into

120

useful objects. Later he learned how to melt these metals
and pour them into special receptacles to solidify into a
desired shape.

We like to think that our culture originated with the
Greeks, but the Greek philosophers took a poor view of
practical inventions. Even the famous Museum of Alexandria,
a kind of national research council, seems to have produced
no really important and useful invention. The Romans were
an intensely practical people, but they seem to have inherited the Greek snobbishness toward inventions and to have
left most of such to the hardworking engineers and artisans.

Invention in the modern sense, using scientific principles, began in the Italian Renaissance, and was closely
associated with art. The age is typified by Leonardo da
Vinci (1452-1519), who devised hundreds of inventions,
which with modern materials and sources of energy would
be quite practical. Unfortunately most of these devices
never got off the pages of his left-handed note books.

Patents

The greatest incentive to invention is financial reward. Even inventors have to eat. In the English-
speaking world the beginning of the patent system was
under Queen Elizabeth I, who granted special monopolies
for the manufacture and sale of devices. But this practice
aroused opposition because of arbitrary allotment. So
in the reign of James I a patent law was introduced. The
word "patent," of course, means open or obvious, and refers
to the fact that the inventor is given a monopoly to
produce and sell in exchange for making his idea public.

The next important step in the evolution of the British
patent occurred under Queen Anne when the requirement was
introduced that the patentee set forth clearly how his idea
could be transformed into a useful object. This statement
was called a specification, a term still in use. Another
improvement came into use under William IV, in which the
inventor "disclaimed" those portions of his device that
were not original with him.

During the eighteenth century and later the British
Government offered special awards for much-needed

inventions. The most famous of these is the Harrison chronometer. In 1707 the great British admiral, Sir Cloudesley Shovell, lost his fleet and ultimately his life on the Scilly Islands because his navigators could not determine longitude accurately. The Admiralty offered a reward of £20,000 for an accurate sea-going chronometer. John Harrison, an English clockmaker, devised a compensating pendulum chronometer that met the requirements. He had almost as much trouble getting the award as in making the invention.

In North America prior to 1783, citizens of the British colonies patented their inventions under the British laws. In 1790 the United States Patent Office was established. Its building and records were spared by the British in 1814, but were destroyed by fire in 1836. Congress took the occasion to pass a new patent law, under which numbers were assigned serially to patents as issued, and the applicant for a patent had to establish clearly the originality of his invention.

The Canadian Patent Office was established in 1824. The first patent was issued to Noah Cushing of Quebec City on June 8th of that year for a "Washing and Fulling Machine." In 1827 a branch of the Patent Office was established for Upper Canada, but it was not until 1831 that it issued its first patent, to Nicol Hugh Baird of Nepean for a method of constructing a wooden suspension bridge. The duration of these early Canadian patents was the same as for British patents, fourteen years. Only British subjects could be granted Canadian patents. After Confederation the Patent Act of 1869 was passed. This limited the period to five years, but opened the right to apply to persons who were not British subjects. Under present Canadian patent regulations the term of a patent is seventeen years.

One problem with patents is that there is no mechanism involved whereby infringement can be prevented in advance. Only by litigation can a patentee obtain legal enforcement of his monopoly for the manufacture and sale of his invention. This has led to the development of corporate research facilities, in which the inventor is employed as one

of a group to develop innovations, which may be patented
by the individual but are assigned to the company. This
device has been employed by universities to derive financial returns from inventions by faculty members. The
institutional approach to invention can be traced to
the establishment in 1876 of the Edison laboratories
at Menlo Park, New Jersey. Here a staff of mechanics
and assistants constructed and tested the devices that
the inventor designed in advance.

Patents as Historical Documents

The Canadian Patent Office, Department of Consumer
and Corporate Affairs, is located at Place du Portage,
Hull, Quebec. Here in the library are preserved the
originals of all Canadian patents beginning in 1824, as
well as those of other British North American colonies
prior to Confederation. Until 1855, summaries of patents
were printed in the series Patents of Canada, but printing
of Canadian patents was then suspended until 1873, when
Volume 1 of the Canadian Patent Office Record was issued,
and this publication has continued. In addition to
Canadian patents, the library here also has a nearly
complete series of the printed claims and specifications
of U.S. patents, as well as those of Great Britain.

The student of Canadian invention will find it necessary to consult foreign patents, especially those of the
United States. The U.S. Patent Office, in Arlington,
Virginia, has in printed form all its patents issued since
1836. Of the "name and date" patents that survived the
fire of 1836, photocopies are available from the U.S.
National Archives and Records Service, Washington, D.C.

The libraries of various Canadian cities have records
of patents. In Toronto the library of the University of
Toronto has a nearly complete series of the Annual Reports,
U.S. Commissioner of Patents, 1847 to 1868. The same
series, 1850 to 1871, is also represented in the Metropoli-
tan Toronto Reference Library, followed by a complete set
of the Index and Gazette. Another good series, 1872 to
the present, is preserved in the Great Hall Library,

Osgoode Hall, Toronto, along with a complete set of the printed Canadian patent records.

Accumulative lists of patents exist, although they cover only part of the record. In 1882 the Canadian Government published the "List of Canadian Patents from the Beginning of the Patent Office, June, 1824, to the 31st of August, 1872." Unfortunately this is long out of print, but it could easily be reissued in replica. Canadian patents subsequent to 1872 are reviewed in the weekly and annual Patent Office Record. The corresponding compendium for the U.S. Patent Office is the three-volume "Subject-matter Index of Patents for Inventions issued by the United States Patent Office from 1790 to 1873, inclusive," compiled and published under the direction of the then Commissioner of Patents, M. D. Leggett, in 1874. The patents are listed under subject, then alphabetically by the name of the inventor. The location of patents by date or number may require search of the Annual Report or Gazette for that year. The Leggett Index was long out of print, but recently it has been reproduced and made available by the Arno Press of New York (1976).

One of the limitations of patents as historical records is that they provide no evidence as to the extent to which the object of the patent was manufactured and used, or even if it were ever produced at all. This gap is best filled by the object itself, especially if it bears the date or number of the patent. Next to the object itself the most helpful record is provided by catalogues of manufacturers and dealers, especially if they are dated. Many such catalogues have been reproduced in recent years, mainly for the use of antique collectors and dealers.

Biographical information on the inventor himself is not provided by the patent, other than his place of residence at the time of issue. I have found that the best source of such information is the community library. Here may be found local yearbooks and civic directories, and an occasional commemorative volume, in which detailed information on prominent citizens or manufactories is set forth.

These libraries often have important clipping collections, as well as unpublished manuscripts of biography or community history.

Canadian Inventions

The most important factor in Canadian invention was the proximity of the United States. During the period from 1824 to 1869, when Canadian patents were granted only to British subjects, Canadians could obtain patents on innovations that had already appeared in U.S. patents. This was deeply resented by American inventors, and the *Scientific American* complained bitterly about the practice. It is probable that in some instances American inventors made arrangements with Canadian citizens to patent their invention under the Canadian's name. Isaiah Jennings of New York obtained a U.S. Patent in 1830 for a lamp fuel consisting of a mixture of ethyl alcohol and turpentine. This came to be known as burning fluid. The following year a Canadian patent (No. 25) was issued to John Ratcliff of Odelltown, Lower Canada, for the same mixture. The wording of Ratcliff's patent is so close to that of Jennings that this was either a case of private arrangement or a very glaring example of invention stealing.

There were no corresponding restrictions on the basis of nationality on the issue of U.S. patents, and Canadians took out such patents, either in addition to a Canadian patent or as their only patent. Abraham Gesner of Nova Scotia in 1846 invented a process for distilling a hydrocarbon lamp fuel from coal or oil shale. He called his product kerosene, but he could not get support for its commercial manufacture in Halifax, so he went to New York and obtained three U.S. patents for the manufacture of "kerosene burning fluid" in 1854. These became the basis for the first petrochemical refinery, which Gesner managed in Brooklyn, N.Y., but the delay lost him his deserved priority over the Scottish chemist James Young and his paraffine-oil (1852). After Gesner returned to Halifax in 1863 he obtained a Nova Scotia patent, but he died in 1864 before he could exploit his invention there.

When the Canadian prohibition against granting patents
to foreign nationals was removed in 1869 there was a rush
of applications by American inventors to obtain Canadian
patents. This was not so much to promote manufacturing
in Canada, as to insure against patent infringement in
this country. In a few cases, however, Americans did move
to Canada and set up manufactories to exploit the Canadian
market and to use patents of doubtful validity where they
were less likely to be challenged. One such entrepreneur
was Richard Mott Wanzer of Buffalo, who established a
sewing-machine factory in Hamilton, Ontario. His machine
was a successful combination of the inventions of Allen B.
Wilson and Isaac M. Singer, and for a time had a wide sale
outside of the United States. Later he purchased the patent
rights to a mechanical kerosene lamp patented by Abel Heath
of New York City in Canada, 1886, and in the U.S., 1887.
Heath's patents were clearly an infringement on patents
obtained by Robert Hitchcock of Watertown, N.Y., in 1872
and 1873, but Wanzer avoided litigation by selling his
lamps in Canada and abroad but not in the United States.

Another American who took advantage of Canadian poten-
tialities was Charles Raymond of Vermont. Raymond was the
patentee and manufacturer of a chain-stitch sewing machine
patented in the U.S. in 1858. In 1862 he moved to Guelph,
Ontario, where he continued to make his machine using the
original design. In 1872 he took out a Canadian patent
for a lock-stitch sewing machine. I suspect that in the
meantime he had acquired Canadian citizenship, for he decor-
ated his later models with engravings of maple leaves and
a beaver.

Some of the most important inventions by Canadians
were made while working in the United States. In 1894 the
electrical engineer Thomas Leopold Willson from Hamilton,
Ontario, was working in Spray, North Carolina, trying to
develop a method of manufacturing metallic calcium by means
of a high-temperature electric current. Instead, he produced calcium carbide, which became the basis of illuminating, welding, and artificial fertilizer industries. After
a few years Willson sold his American patents but retained

his Canadian patents. He returned to Canada and set up
electrochemical plants in St. Catharines, Ontario, and
Shawinigan Falls, Quebec. Later he moved to Ottawa, where
he set up an experimental laboratory, and a pilot plant
at Meach Lake, Quebec. He took out numerous Canadian and
American patents, mostly for processes for making artificial fertilizers. He died in 1915.

Another Canadian who produced his inventions in the
United States was Reginald Fessenden of East Bolton, Quebec.
Working in New Jersey, he made the first wireless transmission by continuous-wave transmission in 1900. In 1902
he obtained a U.S. patent for this type of electromagnetic
propagation and for augmenting or diminishing its amplitude
in accordance with low frequency oscillations such as those
of sound. The first radio broadcast in history was made by
Fessenden in 1906. His continuous-wave oscillations were
produced by a high-frequency electromechanical generator.
Later (1907) the three-element vacuum tube of Lee de
Forest provided a more efficient and practical high-
frequency oscillator. Fessenden's later years were embit-
tered by prolonged litigation with Westinghouse. He died
in 1932.

Taking out a patent for the same invention in the
United States and Canada backfired on one occasion. In
1879 Thomas Edison produced a practical incandescent-
filament electric light, for which he applied for patents
in the United States and Canada. The Canadian patent was
granted on 17 November 1879, whereas his American patent
did not become effective until 27 January 1880. As
Canadian patents under the Act of 1869 had a duration of
only five years, Edison's patent expired in 1884, and
under the then United States law, his U.S. patent was
automatically cancelled as well.

Thomas Edison was the symbolic American inventor,
the poor, self-taught mechanical genius, who could by
himself conceive of new devices and put them into practical
form. But Edison's Menlo Park laboratory laid the foundation for the great corporation research facilities and the
government research councils, where highly trained

scientists, aided by vast support staffs, grind out new
devices, preparations, and techniques faster than industry, much less public demand, can absorb. The individual
inventor, the dedicated mechanic tinkering in his basement
workshop, seems like an extinct species. But there are
still a few of the oldtime creative craftsmen around.
Armand Bombardier with his snowmobile was one of these.
So was Steve Pasjack of Vancouver with his foldaway beer-
carton handle. Let us hope that there will always be a
few such survivors from the glorious age of invention.

COMMENTARY:
J. W. Abrams

I cannot concur with the summary history of invention
given by Russell which treats the medieval period as
sterile. The crank and crankshaft, the fulling mill, the
blast furnace, the treadle, the windmill, the horse collar,
and the assarting ax (which shaped much history) bear
witness otherwise. Lynn White Jr. has given other examples
and Bertrand Gilles has related much on the medieval fore-
runners of Leonardo.

Neither can I accept Russell's heavy emphasis on
economic incentives. As Arnold Pacey has shown, these were
not completely dominant. However, economic incentive is
important and becomes more so with the emergence of a money
economy. Prizes, such as that won by John Harrison (which
he mentions), appear on the scene and the patent system was
introduced to foster innovation and invention. It was
far from perfect and the individual inventor had difficulties in enforcing his rights through the courts.

However, we note from Russell's paper that a patent
system may afford an incentive for an inventor to migrate.
While the patent system may act in this manner, I believe
the general industrial milieu is as important now as it was
during the Industrial Revolution. At that time the general
British situation fostered a Continental "Brain Drain."
Here we note the early cases of brain drain between Canada
and the United States, originally mutual, but later uni-
lateral in the cases of T. L. Willson (who did return to

Canada after selling his U.S. Patents) and that of
Reginald Fessenden (whose work deserves a great deal more
study). Transborder traffic is not unusual in the cases
of inventors, but the reasons deserve study and publiciza-
tion.

Much has been made by our neighbours to the south of
"Yankee Ingenuity" as a characteristic of their culture.
I have wondered for many years whether or not the unique-
ness of this attribute is a self-deception, whether ingenuity *per se* is not distributed more widely and equitably
throughout the world with Canada having its fair share.
We recognize that it was economic factors--not a monopoly
of native ingenuity--that fostered the start of the Industrial Revolution in England. We have evidence that inventors migrate to a favourable environment for development.
We should find out more about what constitutes a favourable
environment--patent laws may be one factor.

Russell has shown us through the examples of Armand
Bombardier and Steve Pasjack that the ingenuity of the
old-time craftsman is still with us and in Canada. We can
see their first actions in Patent Office Records. The
next generation of historians of technology will follow
the progress of their creativity. There are unlearned
lessons to be gained from history, but then the Romans
knew that, too. Can historical knowledge change the future?

DISCUSSION:

The rapporteur, Robert W. Passfield, reports that
the workshop turned from the questions of the patent system
in Canada to the problem of the role of multinational
corporations in the Canadian economy. Many felt that the
very large percentage of industry which is foreign-owned or
-controlled has had a large impact upon Canadian inventions.
Although there seems to be no lack of inventions here, the
lack of risk capital has had an adverse effect upon the
development, manufacture, and marketing of wholly Canadian
products.

There were numerous suggestions as to the type of
studies that might be undertaken to understand the processes

at work in Canada, and some, particularly case studies, might be particularly valuable. It was felt, however, that work along the lines of those suggested by Russell (patent records) would be a valuable beginning.

A.2: ADAPTATION AND/ET INNOVATION

TWO FACTORS IN THE EVOLUTION OF
CANADIAN SCIENCE AND TECHNOLOGY
James O. Petersen

The important preconditions for the development of
science and technology are both social and economic in
nature. Innovation requires a fully developed scientific
and technical culture. Nations that find themselves on
the threshold of industrialization have tended to acquire
their science and technology through a process of adaptation, but even adaptation has its preconditions. Science
could not flourish without scientists, who, in turn, required an organizational framework that would sustain and
coordinate their activities, and an economic role had to be
defined for them so that they could support themselves.
The early organizational framework for Canadian science was
provided by scientific societies and associations of scientific professionals. An economic base was created by
linking science to industry and plowing some of the surplus
produced by the scientifically rationalized industrial
machine back into science itself. Science, in turn, had
to increase the breadth of its technological base in order
to overlap its work with the existing industrial technology.
This led to the rapid development of laboratory science.

The transfer of technology to Canada has been better
documented than the transfer of science. The cod drying
stations that sprang up on the Newfoundland coast in the
sixteenth century were examples of a technology that had
been transplanted *intactu* from northwestern Europe. Between
1600 and 1800, ideas, techniques, and devices poured into
Canada in a steady and increasingly larger stream that cor-
responded to the flow of raw materials out of the country.
By the end of this period it was possible to perceive a
number of distinctive Canadian regional styles in the
choice of tools and methods of work. The social and

131

economic requirements of science were greater and consequently it was less well developed.

The large scale transfer of science to Canada was dependent upon the development of a critical level of socioeconomic complexity, and this in turn fed off the transfer of technology from western Europe. The critical point was reached in the middle of the nineteenth century. It was no accident that Canadian railway engineers were among the midwives of Canadian science. In Canada, as in the United States, the railroads represented one of the most socially important branches of technology, and one that generated administrators and leaders accustomed to utilizing a large staff of professionals, many of whom were trained in technical specialities. Railroad men understood the importance of supporting and developing specialized technical knowledge. Sandford Fleming tells us that he originally conceived of the Canadian Institute as an association of men devoted to the technical professions,[1] but that such an association could only survive by redefining itself as a scientific society and opening its doors to all gentlemen with scientific interests. Railwaymen like Alfred Brunel, Frederick Cumberland, and Sandford Fleming provided the financial and administrative nucleus for the formation of a scientific society in which professional scientists like Henry Croft could take root and grow.

Another cluster of scientific harbingers can be found among the technically trained army officers of the British Army who were stationed in Canada during the first half of the nineteenth century. J. H. Lefroy, a captain in the Royal Artillery, organized and operated the Toronto Magnetic Observatory for two decades before turning it over to J. B. Cherriman, a professor of science at the University of Toronto.

The social relations between scientists and technically trained professionals were critical for the beginnings of science in Canada, and scientific societies provided a milieu in which these relations could be strengthened and extended. But the importance of the Toronto Magnetic Observatory in the early history of Canadian science

should remind us that science, like industry, required an
increasingly higher level of technology in order to continue to grow. With the achievement of a network of
scientific societies and the consolidation and definition
of the social relations of science in Canada, attention was
gradually shifted to the problem of acquiring the "means
of production" of science. Henry Croft achieved a major
step forward in this direction when he succeeded in financing, constructing, and equipping the chemistry laboratory
at University College in Toronto in 1856.

In this respect, the history of Canadian science and
the history of Canadian technology intersect. Both science
and technology in Canada developed from a base of mature
technology imported from the main centres of the industrial
revolution. The adoption of laboratory technology and its
adaptation to the solution of Canadian problems was an important event in the history of Canadian science. The
first laboratories were established for the purpose of providing instruction in the physical sciences. They were
intended to be used as instruments in the transfer of
scientific ideas to Canada. In this respect, laboratories
were rather like the museums that were being set up at
the same time. They were the physical manifestations of
a system of ideas generated in another time and place.

Adaptation could only take place after the laboratory
had been moved from the sphere of ideas to that of practice.
The necessary condition for the generation of a first
approximation to Canadian science--a Canadian adaptation of
European science--was its integration with production and
its grounding in the material realities of Canadian life.
Laboratories were established at Queen's University for
the purpose of pursuing industrial research--chiefly in
connection with the mining industry--in the eighteen-
nineties, and perhaps we can see here, once again, the
guiding hand of Sandford Fleming. The University of
Toronto Physics Laboratory had been established for research purposes by the decade of the First World War,
while the University of Toronto Physiology Laboratory had
been established for research purposes by Thomas Brodie a

few years before. Only after the accumulation of a substantial technological base, and the training of an entire
generation of research scientists in the techniques of the
various branches of laboratory science, did innovation become a real possibility in Canadian science. The history
of scientific laboratories is, therefore, an important part
of the history of Canadian science, and the biographies of
laboratories are as important as the biographies of scientists.

It is important to note that many, if not most, of
the early labs in Canada were set up by the state, rather
than by the private sector or by the universities. In
this respect, developments in Canada more closely paralleled those in Germany, where the state was a vital factor in the development of science, than those in England
where the state was less visible. Like the German state,
the Canadian state was confronted with the problem of con-
verting a rural agricultural nation into a modern industrial one. The state was the only body in Canada with
access to sufficient capital to employ science on a significant scale, but the development of the state laboratories that played such an important role in Canadian
science in the twentieth century was as much the result of
accident as it was of planning. In the brief time remaining
I would like to examine the factors that contributed to the
formation of the laboratory of the federal Department of
Inland Revenue, as an example of the development of a
state laboratory.

The Department of Inland Revenue was created after
Confederation for the purpose of collecting excise taxes,
canal tolls, and a few other miscellaneous sources of
state income. It had taken over some of the functions of
the old Customs Department of the government of the United
Province of Canada. Chief among the sources of income was
the very high excise tax on distilled spirits, malt liquor,
malt, and tobacco, which had been imposed by the previous
government in 1864. The rates of taxation, in some cases
double the cost of production, encouraged evasion and required strict supervision of a technologically dense and

capitally intensive industry. Technical competence was
an important desideratum in the inspectors called upon to
enforce the law in the name of the department. To make
matters more complicated, the department was also charged
with collecting an excise tax from the distillers of petroleum.[2]

By 1871, the Department of Inland Revenue had become
a major source of income for the state, and its functioning
was too important to be left in the hands of politicians.
A. Brunel was appointed Commissioner, presumably on the
strength of his understanding of canals, and in antici-
pation of his ability to extract more income from them, but
he took a lively interest in all aspects of the work,
which increased in scope and complexity as each new year
saw the passage of new laws to be enforced by the department. In 1870, the department was required to inspect
the production of vinegar and methylated spirits and to
collect a duty from them. In 1871, the excise on many
petroleum and coal tar distillates was removed, but, in
1873, the department was given the responsibility for in-
specting staple items like butter, flour, beef and pork,
leather and potash. In addition it was charged with in-
specting the weights and measures of the illuminating gas
coming into use in the cities, to assess its powers of
illumination and to test it for impurities like hydrogen
sulphide.

Clearly there was a major conjuncture between two
traditional functions of the state--taxation and regula-
tion--and the rapid growth of science and technology. In
1875, the Canadian government permitted brewers to reclaim
the excise tax on the beer they exported and provided that
the refunds were to be calculated in proportion to the
amount of malt contained in the beer. At first, the
brewers were allowed to make their own claims, but the
claims were so ridiculously overstated that the Department
of Inland Revenue set about devising some laboratory tests
that would enable it to determine the amount of malt that
had been used to produce a given amount of a particular
beer. Brunel's idea was to use a series of tests that had

been adopted in England, but he had to prepare for the
eventuality that these tests would be legally challenged
as not being applicable to Canadian conditions. He was
therefore forced to make a Canadian adaptation of British
laboratory practice in response to a conjuncture of legal
and administrative problems.[3] Brunel was quick to realize
the significance of this and he argued for the formation
of a technically competent body of civil servants, selected
on the basis of examination, as provided in an Order in
Council of 1866.

> The equitable charge of these important
> taxes largely depends on the intelligence
> and technical knowledge of the collectors
> and their subordinate officers, and it is
> therefore, obviously necessary, for the
> protection of these manufacturers, as well
> as for the safety of the revenue, that these
> officers should be thoroughly competent, not
> only as mere clerks, but as regards such
> technical knowledge as is required of them.[4]

Brunel insisted that his inspectors demonstrate competence in computing volumes, measuring surfaces, using a
hydrometer and a slide rule, and that they possess a
general knowledge of distillation, brewing, or the production of tobacco. He became a conscientious agent in the
restructuring of the state bureaucracy to accommodate the
new scientific culture. He introduced new kinds of
social relations into the bureaucracy.

Brunel was doing for the Department of Inland Revenue
what Logan was doing for the Canadian Geological Survey,
or what DeLaBeche had done for the British Geological Survey, except that he was operating within a less well-defined
legal framework. In each of these three cases, we have the
opportunity of studying the articulation of science onto
the state, and there are others of equal importance that
could have been chosen. Finances were an important factor
in motivating the state, and revenue from the taxation of
technologically dense industries was a logical starting
point for the growth of science within government. Sometimes the pressure would come from outside, as when the
Royal Society lobbied for the formation of the Biological
Board, or when the sugar industry lent its voice to the

demand for the formation of the National Research Council.

Whatever the source of the pressure on the government, whether internal or external, the results were the growth of social base composed of technically trained civil servants and a technical base of increasingly complex and expensive laboratory equipment. The large costs of these two factors of scientific production almost guaranteed that in Canada, the state would have to play a very important role. The two factors tended to develop simultaneously. In 1874, at the same time that Brunel was recruiting his technical gentlemen to work as inspectors for the department, he also received a photometer, and a chemical laboratory designed to detect impurities in gas.[5] The chemical laboratory was destined to serve the limited purpose of regulating the gas industry, but it was probably one of the most sophisticated Canadian facilities of its time.

In many ways, the transfer of modern science to Canada has paralleled the transfer of modern technology. Both were foreign imports that had to be adapted to Canadian conditions. In both cases, the structure of Canadian science had to be altered before the transfer was successful. Both came as mature systems from older and more developed centres. Adaptation and innovation were processes that were contingent upon successful transfer. Those sciences that were successfully transferred, like geology, were in some sense preadapted to the Canadian environment. Some modes of perception were more easily assimilated than others. The natural historical view of science found easy acceptance in Canada, while laboratory science, with its close links to industrialism and its dependency on refined technology and highly developed technical skill, posed many problems. It is not clear, in retrospect, whether laboratory science was adapted to Canada or Canada to laboratory science. In any case, the adaptations of the pre-First World War period were the precondition of whatever innovations were possible in the postwar period. The historian of science can perceive a chain of events beginning with a series of transfers, leading through a difficult period of

138

adaptation based, all too often, on simple trial and error, and leading to a period when it is possible to recognize some genuinely innovative work which signals the birth of Canadian science as a thing in itself.

NOTES

1. Sandford Fleming, "Note on the early days of the Canadian Institute," and "Early days of the Canadian Institute," in Canadian Institute, *Transactions*, vi, (1898-99), 1-657.

2. Alexander Morris, "Report of the Minister of Inland Revenue for the fiscal year ending 30th June, 1870," in Canada, *Sessional Papers*, 3, no. 6 (1871), 1-11, and also Morris' report for 1871 in Canada, *Sessional Papers*, 4, v, no. 6 (1872), 1-26.

3. Alfred Brunel, "Memorandum respecting the payment of drawback duty on malt contained in beer exported," in Canada, *Sessional Papers*,10, iii, no. 4, xxxi-xxxii, and "Report of the Commissioner of Inland Revenue for the Fiscal Year ended 30th June, 1876," in Canada, *Sessional Papers*,10, iii, no. 4, xviii.

4. Alfred Brunel, "Report of the Commissioner of Inland Revenue for the fiscal year ended 30th June, 1874, in Canada, *Sessional Papers*, 8, ii, no. 2 (1875), xxv.

5. *Ibid.*

COMMENTARY:

George Sinclair

I contend that the problems facing the historiographers of technology stem not so much from what they are doing as from what they are not doing. Specifically, the deficiencies in the discipline are due to our inadequate perception of what is the nature of *engineering*. Unfortunately, the historiographers can expect no help from the engineering community, as the engineers cling to obsolete concepts of what constitutes engineering, which, in turn, can be blamed on the lack of a philosophy of engineering.

The critical issue is the matter of the *human involvement* in the creation of the works (artifacts) of engineering. It is quite astonishing, for example, to read a typical history of technology[1] and discover that, although it deals mostly with *engineering* developments, there is no concern whatsoever with the human participation, beyond

quoting the names of the engineers. The history of technology is a history of gadgets and devices. It is no
wonder that Jacques Ellul claims that the new technical
milieu is artificial, autonomous with respect to values,
ideas, and the state, and is self-determinative *independently of all human intervention.*[2]

The importance of the human involvement has, of
course, not escaped the attention of historians. John B.
Rae, when he became President of the Society for the
History of Technology, chose as the title for his presi-
dential address "Engineers Are People."[3] The historians
have only partially grasped the important fact that the
major problems of society are basically created by humans
and will only be solved by humans.

It should be evident that it is not sufficient just
to observe that "engineers are people." The *nature* of the
human participation is the key issue, and this is a question
which has been virtually ignored. It is my thesis that we
need a *philosophy of engineering*, or more properly, a philosophy of the *professions*, to provide the quidelines needed
by historiographers of technology if they are to integrate
the human element into their histories, and if they are to
make their discipline have relevance to the human problems
of modern society.

We need more sophisticated concepts of what are the
characteristics associated with individuals who exemplify
the best in any given profession. My suggestion is that
there are three characteristics which distinguish the professional from individuals in other vocations:

(1) the courage to create or to lead,

(2) a deep *personal commitment* to the adequacy of the
    advice given to clients, while recognizing the
    small possibility of an error, and

(3) *maturity of judgment*, based on experience, special-
    ized knowledge and continuing education.

It is not generally recognized that creativity is the
primary attribute of the true professional in any field, nor
is it understood that it takes considerable courage to be
creative, to be a leader.[4] Courage is required in order for

the professional to maintain a deep and personal commitment to a decision, once it has been made, while recognizing the finite possibility of there being an error in
the decision resulting from the inadequacy of the available data.

With reference to Dr. Petersen's paper, it is my
opinion that to focus on the "transfer of science and
technology" is to ascribe undue importance to the transfer
process. The crucial factor is the human involvement,
because the significant history has to be the history of
people and persons, not the history of things and processes.
As Dr. Petersen points out, the social relations between
scientists and technically trained professionals were
critical for the beginnings of science in Canada. I
would like to have seen this theme developed.

To my mind, the historical significance of Henry
Croft's acquisition of the "means of production" of science
was in his courage and commitment to the creation of the
laboratory rather than the act of importing apparatus. The
apparatus was available to many Canadians but it was Croft
who provided the *leadership*. I have difficulty in accepting
Petersen's statement that the "biographies of laboratories
are as important as the biographies of scientists."

The historian of technology labours under a severe
handicap, namely, the lack of guidelines to indicate what
important factors need to be examined. This is due to the
considerable ambiguities associated with the term "technology." As Langdon Winner[5] (in a review of a volume called
Research in Philosophy and Technology, vol. 1, 1978) has
pointed out, the best that one can say of the philosophy
of technology is that there isn't any. The philosophy
which the historiographer requires is not of technology
but of the professions.[6]

NOTES

1. Melvin Kranzberg and Carroll W. Pursell Jr., *Technology
    in Western Civilization* (Oxford: Oxford University
    Press, 1967), 2 volumes.

2. Carl Mitcham and Robert Mackey, *Philosophy and Tech-
    nology* (New York: The Free Press, 1972), 86.

141

3. John B. Rae, "Engineers are People," *Technology and Culture* 16:3 (July 1975), 404-18.

4. Rollo May, *The Courage to Create* (New York: Bantam Books).

5. Langdon Winner, "Philosophers on Technology," *Science* 202 (6 October 1978), 44-5.

6. George Sinclair, "A Call for a Philosophy of Engineering," *Technology and Culture* 18:4 (October 1977), 685-9.

DISCUSSION

Rapporteur David F. Walker reports that there was a great deal of methodological debate touching upon such issues as the meaning of the word 'technology,' whether knowledge of the brain's functions in creativity is useful to the historian, and whether developing a philosophy of engineering is worthwhile.

Issues of a more practical nature brought discussion to pleas for better (and more) oral history projects on innovation, suggestions for a study of the barriers to innovation, and a strong desire to broaden the archival base for the study of innovation (i.e., to encourage companies and professional societies to maintain archival collections and open them to the historian).

A.3: INSTITUTIONAL FRAMEWORK/CADRE INSTITUTIONNEL

THE STUDY OF THE INSTITUTIONAL FRAMEWORK
FOR SCIENCE AND TECHNOLOGY
J. W. Grove

It is not uncommon for presenters of conference papers
to begin by saying that their topic was not of their
choosing but was assigned to them by the organizers. I say
this with particular poignancy on this occasion because I
have to admit at the outset that I am far from sure that I
have divined their intentions correctly and would much
rather have written about something else! I have assumed,
however, that we have not come to this workshop to talk
only, or even at all, about the history of "science policy,"
because it is only in very recent times that the notion of
a fully explicit, overall "science policy," to be pursued
in concert by those who *do* science and technology and those
who only prescribe what they are to do "in the national
interest," has been conceived in *any* science country.
"Science policy" in this sense seems to be far too recent
and altogether too elusive for adequate historical study.
No doubt someday it will become the proper subject for an
historian's attention.

If, on the other hand, by the history of "science
policy" we understand the history of state sponsorship,
direction and promotion of science and technology for
societal ends, we are talking of something not only more
tangible but much broader in scope. We are talking about
the history of a policy process, or perhaps one should say,
about the histories of a succession of policy processes.
And a policy process is not something that governments
alone participate in--it is a process in which many different institutional actors play a role.

My second assumption was that the object of the workshop is not to look at substantive history, but at research
programme and method, i.e., what kind of work might in

142

principle be done and how much of it is feasible given
the available resources and the interests of scholars.
What little I may have to contribute to the discussion
is necessarily coloured by the fact that I am not an historian, but a political scientist who professes, among
other things, Public Administration. It follows that the
kinds of issues that are likely to interest me are not
necessarily those to which historians of science and technology would attach the greatest importance. Since
the paper *is* prepared for a workshop, I have assumed also
that you would wish me to keep it short.

I should add one further disclaimer, and that is that
I have balked at the idea of interpreting "science and
technology" (henceforth ST) as equivalent to "R & D" (an
expression that materialized, to my alarm, in an earlier
version of the programme), partly because to get into the
development end of the spectrum would seem to make our task
today utterly unmanageable, but mainly because I know
nothing about "D"--which I imagine you will accept as a
sufficient reason for my not doing so.[1] So--by "ST" I
mean, roughly, pure and applied science.[2]

Let us glance, then, briefly, *at* "the institutional
framework for Canadian ST." We can envisage this as consisting of four potentially interacting sets of federal
and provincial agencies: the Controllers, the Providers,
the Doers, and the Influence Seekers.[3]

The Controllers are the agencies that have developed,
over time, responsibility for resource allocation and/or
some measure of policy-setting and central direction in
respect to ST. Some of these agencies are "global" (in
that their actions may affect ST as a whole): the Cabinet
and the central machinery of government (in contemporary
terms, Privy Council Office, Treasury Board, Ministry of
Finance, etc.). Others are specific to areas of ST, e.g.,
the Atomic Energy Control Board, government departments
manifesting CD/PS/RA functions in respect to, e.g., forestry
research, agricultural research, etc.

The Providers are the agencies providing actual,
direct support in money and material for ST: granting and

contracting agencies in both the public and the private
sectors--in broad terms, the patrons of ST.

The *Doers* are those who actually do the research in
ST either as the whole or part of their function, including:

(1) governmental and quasi-governmental agencies,
i.e. (in contemporary terms), Department of
Energy, Mines, and Resources, Atomic Energy of
Canada Ltd., etc.

(2) quasi-public agencies such as university science
and engineering departments and (some) research
institutes.

(3) non-public agencies, e.g., industrial firms, trade
associations, (some) research institutes.

The *Influence Seekers* are all those who may seek to
influence the controllers and/or the providers. They are:

(1) "governmental," e.g., Parliament and its committees; bodies such as the Economic Council of
Canada and the Science Council, and

(2) "extra-governmental," including:

(a) scientific societies and professional bodies,
which may be either "global," e.g., The Royal
Society of Canada, SCITEC, ACFAS, being concerned with ST as a whole"; or "sectoral,"
being concerned with the advancement of a
single "discipline" or the protection and regulation of a particular profession (and sometimes both), e.g., The Entomological Society
of Canada, the Canadian Association of
Physicists

(b) trade associations and other industrial
interest groups

(c) mainly more recently (but I am sure there must
be interesting examples from the remoter past)
what are now increasingly called "public-
interest science" groups, e.g., environmen-
talists, anti-nuclear groups, etc.

These are not, of course, completely exclusive categories: for example, the doers themselves may, from time
to time, attempt to join the ranks of the influence seekers;
and the providers indirectly influence the doers by directly
allocating resources through the granting process.

This "map" of the actors that constitute the institutional framework of ST is, of course, not static. It is
continually dissolving and reappearing over time with its

details (though not its general configuration) changed.
Equally obviously, ST is not a monolithic entity (though
no doubt for general histories it is possible to treat
"science" or "technology" or even "science and technology"
as if they were) but rather something that can be "sec-
torized" in various ways, e.g., into "disciplines"
(entomology, earth sciences), "programmes" (forestry policy, mineral resource policy), or major technologies. Moreover, both S *and* T are themselves changing. They are
changing not only cognitively (that is, as bodies of knowledge) and in the sophistication of their techniques but
also systemically (that is, in their internal social
structure and in their interactions with the environment
constituted by the state and society). Technology, as we
know, has developed from a craft and skill base (knowing
how without knowing why) to a science base (knowing how because knowing why) to become *systematic* technology: technology (or rather, technologies) founded on the application of science and organized and cultivated like science,
and systematic too in that changes in one technology have
come to require changes in other technologies. They have
become increasingly professionalized, specialized, and
organized on a large scale.

Science shares these characteristics. The rise of
"Big Science" has led to changes in the work situation of
scientists: team research, the large-scale research institute (often "mission-oriented"), complex and very expensive
equipment--the phenomenon predicted, with characteristic
foresight, by Max Weber, of the workplace of science as a
factory, with the individual scientist increasingly aliena-
ted from the means of production and directed as to how
the means are to be used by administrators of science. The
picture of the rising importance of Big Science as charac-
terizing and dominating *all* science may well be overdrawn,
it is true , because the bulk of Western sociological and
politico-administrative writing about science is American,
reflecting the dominance of American science since World
War II (at least half of all qualified scientists and
engineers--QSE--in the world today are American--or at any

rate resident in the United States--and most of the rest
are in the USSR); but this merely increases the interest
and utility of taking American experience as a bench mark
for the study of the recent history of ST in "Little
Science" countries like Canada, with the added interest,
in the Canadian case, of exploring what impact science, as
done and organized south of the border, may have had on
Canadian ST.

Moreover, there are other characteristics of the
developing structure of science that apply (though perhaps
to a lesser degree) as well to Little Science as they do
to Big Science. Among these, importantly, are its increasing bureaucratization and politicization. These are to
some extent interrelated. The degree of politicization of
science is a function of the extent to which science is
seen to be essential to technological development and the
degree to which both are seen to be essential to the pursuit of national goals. Some of the bureaucratization of
science (e.g., an elite hierarchy attendant on the growth
of the science advisory system) results from politicization.
On the other hand, some of the politicization of science
can be related to its bureaucratization (e.g., largeness
of scale means escalating costs to be met from public
funds, which increases the demand for political control
in the interests of public accountability).

Much attention is now devoted by historians and
sociologists to systemic change in science, at least inso-
far as concerns its purely internal structure (the reward
or exchange-of-recognition system, the reception-of-ideas
system, etc.), but, while the importance of "externalist
history (and sociology)--i.e. history and sociology that
seeks to relate internal change to the influence of social
forces--is now widely stressed, rather little attention has
been given to investigating the impact on internal structure
of the wider *institutional* framework: for example, how the
flow of money and scientific manpower, increasingly under
the control and manipulation of governments (the controllers
and public providers), may affect the system of social
relations within science itself.

Thus, the picture of the institutional framework for ST, I suggest, looks something like this:

```
↑                                                    ↑
Institutions and              Cognitive and internal social
institutional mix             systems of ST "disciplines"
changing over time            changing over time

┌─────────────────────────────────────────────────────┐
│                                                       │
│                          Sectors* of ST              │
│                                                       │
│              Controllers   X    X    X    X    X      │
│                                                       │
│              Providers     X    X    X    X    X      │
│   ACTORS                                              │
│              Doers         X    X    X    X    X      │
│                                                       │
│              Influence     X    X    X    X    X      │
│              Seekers                                  │
│                                                       │
│   *  Disciplines," "Programmes," "Technologies," etc. │
│                                                       │
└─────────────────────────────────────────────────────┘
```

The institutional framework consists of the interacting actors. But they act only in relation to identifiable "sectors," which may be "disciplines," "programmes" (or policies), specific "technologies," etc., which the researcher selects for study. The focus of a study would be on a process in which some or all of the actors are involved in respect to a sector or sectors (i.e., one or more related programmes, disciplines, or technologies). Depending on the time span selected by the researcher, or imposed by the nature of the subject chosen for study, both institutional change and cognitive and social systemic changes may be relevant.

Merely to "map" the institutional framework in this
way is to construct a formidable research programme and to
open up--in principle--an immense number of possibilities
for historical research: "in principle," because what can
be done will obviously depend upon the availability of
interested scholars, on the richness of the available
primary sources, and on the possibility of securing adequate access to them. These constraints will, in themselves, go a long way towards determining the choice of
subjects; but since even the supply of *material* is likely
to outstrip by several powers the number of scholars available to study it, priorities would have to be determined.
Important among these, I would suggest, would be the
utility of a given study for policy-making and, in particular, for those entrusted with science policy decisions
and resource allocation to science and technology institutions, programmes, disciplines, and technologies.

My own inclinations as a political scientist not un-
naturally tend towards historical studies that proceed
from the perspective of, and with the emphasis upon,
government rather than on what is being governed: the
history of public agencies and their interactions in the
pursuit of programmes, that is, in the process of producing
(and, as often, failing to produce) sets of intended out-
comes--in other words, *administrative* history. But, since
administration is a part of politics, is not carried on
in a vacuum but in a changing societal context, and involves many kinds of actors, not merely administrative
agencies in the narrow sense, I mean administrative history
broadly conceived. It is generally a lack of these wider
perspectives that has helped to make much administrative
history so dull. If we are looking for a model administrative history broadly conceived, we have one in the manner
I intend here in Mrs. Margaret Gowing's superb account of
British atomic energy, 1939 to 1952.

NOTES

1. It is said that Frederick the Great, when visiting some
    town, asked the mayor to explain why no Royal Salute
    had been fired. "There are fifteen reasons why,

sire," the mayor is supposed to have replied, "the first being that we have no cannon." "In that case," said Frederick, "I do not wish to hear the other fourteen."

2. I am unrepentant about using the formulation "pure science" because "basic" implies basic *to* something and "fundamental" means a foundation *for* something, and I strongly reject the now fashionable assumption that the only reason for doing science is, or should be, that it leads to some technological payoff (though I admit that pure science is often done to support a "mission"). It is worth noting that the Russians, who, in accordance with good marxist-leninist principles, make precisely this assumption, *do* talk about "basic" research, while rejecting "pure" science as a bourgeois illusion.

3. In giving (contemporary) examples I have named federal agencies only. Of course, the interactions may be not only between types of actor, but also between federal and provincial actors *within* a type.

COMMENTAIRE:

Brigitte Schroeder-Gudehus

De futures générations verront-elles dans la "science policy à la OCDE" un ensemble de vaines tentatives et d'espoirs futiles?--ou même, à la limite, une tendance néfaste pour la science, dans la prolifération de minis-tères, comités et organismes publics de toutes sortes, tous préoccupés de politique tant pour la science que par la science? Certes, la conviction selon laquelle une organisation plus ou moins complexe de la production du savoir et de son utilisation permettrait d'en optimiser les résultats, attend toujours une confirmation. Toujours est-il que les gouvernements n'ont pas attendu l'ère de l'Organisation de coopération et de développement économ-iques pour intervenir consciemment dans la gestion des ressources scientifiques et techniques nationales, et qu'on ne saurait ainsi analyser et comprendre l'histoire des sciences et des techniques aux 19e et 20e siècles sans tenir compte du cadre institutionnel a l'intérieur duquel elles se sont développées.

Parlant de cadre institutionnel, on aurait tort cepen-dant de réserver ce terme aux seules institutions pub-liques. Une telle approche serait à peine acceptable pour étudier les développements scientifiques et techniques de

l'époque strictement contemporaine, dont une des caractéristiques est précisément l'omniprésence des pouvoirs pub-
lics--dans tous les secteurs d'activité d'ailleurs. Il
convient donc d'étudier ce "cadre institutionnel" dans
une acception plus large, ce qui inclut: les universités
qu'elles soient des organismes publics, comme dans beaucoup
de pays de l'Europe, ou privés comme il est généralement
le cas en Amérique du Nord; les institutions de recherche
extra-universitaires, centres à but non-lucratif ou labora-
toires industriels, par exemple.

Ceci n'est pas suffisant cependant. L'étude doit
aussi s'étendre au processus d'institutionnalisation du
savoir: à l'émergence et la consolidation des disciplines
et des spécialisations, aux systèmes de sanctions se
rattachant à la qualification, et à la professionnalisation des activités scientifiques et techniques qu'ils ont
permise. Il faut s'intéresser, en d'autres termes, aux
sociétés savantes, aux curricula universitaires, aux
examens, aux diplômes, aux ordres professionnels....

Les interactions des divers éléments du "système
scientifique" ne se résument donc pas à un rapport simple
entre les pouvoirs publics d'une part, et les praticiens
de la science et de la technique de l'autre, entre ceux qui
gouvernent et ceux qui sont gouvernés. Je me rallie à la
façon de voir de J. Grove qui fait éclater ces catégories
simplistes et appréhende la réalité par une nomenclature
plus subtile, distinguant entre "controllers," "providers,"
"doers" et "influence seekers." J'avoue éprouver quelques
hésitations à aller aussi loin que J. Grove dans l'atténu-
ation des démarcations entre le secteur public et le secteur
privé: trouver le Parlement et ses commissions réper-
toriés parmi les "influence seekers," répartis--comme il
s'agissait là d'une distinction banale--entre "governmental," me cause un certain malaise. Les représentants du
peuple souverain, ne recherchent-ils pas à exercer une
influence sur la vie nationale en vertu d'une légitimité
très particulière, toute différente de la raison d'être de
sociétés professionnelles ou de groupes d'intérêts industriels?

Or, il ne suffit pas d'englober ainsi dans l'étude du cadre institutionnel les éléments de ce qu'il est convenu d'appeler "le système social de la science," si l'on ne l'étend pas également aux conceptions fondamentales--à l'idéologie, aux valeurs--qui s'expriment tant dans l'organisation de la société que dans le comportement des individus. Il faut s'interroger ainsi sur l'idéologie dominante et la place qu'elle accorde généralement à l'autorité publique et à ses institutions; sur l'importance qu'une société attribue aux activités intellectuelles, sur le rôle qu'elle assigne à l'activité scientifique et technique--bien culturel et/ou facteur de puissance; sur l'estime dans lequel elle tient, par conséquent, ceux qui oeuvrent dans le domaine scientifique et technique. Le fait de détenir un diplôme universitaire, d'être chercheur scientifique ou ingénieur, devient ainsi un facteur de stratification sociale: cela aussi fait partie du "cadre institutionnel."

Il est certain, de plus, que le cadre institutionnel du développement scientifique et technique ne se construit nullepart en vase close, c'est-à-dire en fonction des seules données de la situation nationale. Les exemples abondent d'influences étrangères ayant marqué, dans certains pays, l'évolution de l'enseignement supérieur, l'articulation des efforts de recherche sur les besoins socio-économiques, les mécanismes de coordination et de financement, etc. Ainsi, la création du Conseil national de recherche du Canada faisait-elle partie du "mouvement des Conseils nationaux de recherche," politique inspirée par la conviction que la supériorité scientifique et technique de l'Allemagne, donc sa puissance, devait beaucoup à l'organisation que ce pays avait su donner à son enseignement supérieur, à sa recherche scientifique et à son développement technique. L'identification des influences, de leurs origines et des voies de transmission, l'évaluation nuancée, aussi, de leur portée, ouvre des voies de recherche intéressantes et à peine défrichées.

Est-il alors nécessaire encore de souligner l'importance décisive des sources, d'archives publiques d'abord,

certes, mais aussi d'archives privées? Car cette perception du rôle de la science dans la société, l'influence
éventuelle de modèles étrangers,--bref, tout ce qui fait
partie de l'univers de ceux qui sont gouvernés plutôt
que de ceux qui gouvernent (et auxquels, à mon avis,
J. Grove accorde une attention trop exclusive)--ne peut
être repéré qu'en marge des documents officiels, dans les
notes personnelles, dans les correspondances, dans des
publications éphémères. Nous sommes convaincu que le
Canada est riche en fonds d'archives d'hommes de science
et d'ingénieurs. Encore faut-il les repérer, les conserver
et les rendre accessibles.

DISCUSSION

Louise Dandurand, the rapporteur, brought to the
attention of the conference participants the different archival requirements for those studying the interaction of
science, technology, and society by either of the methods
espoused in the paper and commentary. Grove's "actor"
model, which places great emphasis upon governmental and
paragovernmental bodies, requires public archives of
governmental papers. On the other hand, Schroeder-Gudehus'
view that one must take a global approach demands far
wider resources: not only governmental papers, but also
a variety of institutional and private papers, and sources
beyond Canada.

What seems to be a rather narrowly focussed discussion
here has far wider ramifications. The methodology outlined
by Grove is essentially a sociological model, which can
be utilized for the recent past with little difficulty,
but, as with all sociological models, runs into greater
difficulties as one proceeds further back into time, has
less data, and must face social forces that may not be
strictly analogous to today's. The great advantage to this
approach is its simplicity; it offers one a first approximation picture quickly. Since science and technology have
had such strong interaction with government in Canada,
this method could yield a valuable collection of insights.

Schroeder-Gudehus offers a much more complex concep-
tion--it is not a model, really--which has strong

affinities to the General Systems approach. One of us (Jarrell) believes that this approach is potentially the most powerful of the recent methodologies available to historians since it attempts to deal with very complex relationships. This is a distinction that Schroeder-Gudehus points out: since the world is complex, any model attempting to explain it will have to be complex, at least in its details. But, for the working historian (who is going to distort reality anyway, no matter what methodology he chooses), there is the choice between the workable model that will yield results, or a more "accurate" model that may be virtually impossible to utilize. This, then, is the dilemma that seems to have been left unaddressed by the workshop. Historians, like scientists, know the world is complex, but chop it into small, workable bits, with the hope that someone may, someday, synthesize it all. Glance at the bibliography at the end of this volume and see how the microscopic approach has been the norm. Both Grove and Shroeder-Gudehus are to be congratulated for thinking on a larger scale; they offer us methodologies. What we require now are explicit theories of history and of society. We use such theories in our work, but who has stopped to enquire what they are?

A.4: TEACHING MATERIALS/MATERIEL DIDACTIQUE

AN ASSESSMENT OF TEACHING MATERIALS IN THE
HISTORY OF TECHNOLOGY IN CANADA

George Richardson

This paper deals primarily with teaching materials
for undergraduate courses in the History of Technology in
Canada. Contrary to the statement in the program, this
paper does not cover teaching materials in the History
of Science in Canada. The material covered will be primary
and secondary sources in periodicals and monographs along
with artifacts and field sites. Other materials available
in archives and museums are covered in other papers in the
conference.

The most important point to make is that there is no
single source available that could be recommended as a
comprehensive text for a course. No one has yet produced
a complete survey of the various technologies found in
the development of Canada including an assessment of their
importance and an analysis of their impact on each other
and on the social, economic, political, and cultural life
of Canada. Therefore, each teacher must rely on his own
judgment to plan his course, although some basic principles
must be considered.

The first is that the History of Technology in Canada
must be, above all else, good history and an integral part
of the overall evolution of Canadian society. It must not
be considered in isolation or in any way divorced from
the mainstream of Canadian life. The second is that the
role that technology plays in Canadian society must be
very carefully and accurately analyzed so that one avoids
merely a glorification of technology, a search for superla-
tives, or a binge of nationalistic fervour. The fact that
technology is a vital component in Canadian life cannot be
denied; what is important is to portray its role as clearly
and sympathetically as possible. A sympathetic treatment

is essential because a student only begins to understand technology when he begins to develop a real feeling for it.

The nature of our schools and universities and the interdisciplinary nature of the subject create two types of students. The first has some knowledge of Canadian history and needs to be introduced to the development of technology. Hopefully this student can adequately weave the two together without further help from the teacher. The second may be an engineering student or any student without a background in Canadian history. This student should not be given the history of technology without a nontechnical framework or at least references to material that will fill the gap.

Most primary and secondary sources now available do not adequately relate technology and society; this places the onus on the teacher to compensate. One of the best basic sources for background is Easterbrooke and Aitken, *Canadian Economic History*. This book is reliable, readable, and readily available. An older source is Lower and Innis' *Settlement and the Forest and Mining Frontier*, which serves as a good example for the new student of a reasonable blend of technical and economic history, although the subject matter is somewhat restricted. Nelles' *Politics of Development* is a more modern account of the interaction between technology, politics, and economics and provides excellent background for a general study, although also restricted in location and subject matter. Finally, all students should be made aware of *Let Us Be Honest and Modest*, edited by Bruce Sinclair and others, which introduces the student to certain aspects of technology and the early Canadians' attitude towards it. It provides an atmosphere that is difficult to convey in any other way.

Written information about technology and technological processes will be discussed under two headings, primary and secondary sources. While most primary sources are located in our archives, an amazing amount of information is available in government documents which range from census reports and departmental reports to reports of Royal

Commissions. Many departments such as Public Works, Railways and Canals, and Transport had sections headed by chief engineers or other technical personnel who were required to make annual reports. These can be found in sessional papers or departmental reports. George Henderson's *Federal Royal Commissions in Canada 1867-1966* is a valuable checklist. The preamble to most of these reports gives useful histories that are generally reliable. Provincial Royal Commissions are much more difficult to locate but just as useful. Finally, government documents yield invaluable statistical information.

Early periodicals also contain primary research material. In addition to numerous newspapers, national magazines, and farm journals, several technical publications are available in many libraries. Examples are: *The Transactions of the Canadian Society of Civil Engineers, The Canadian Mining Journal, Contract Record and Engineering News, Railway and Marine World, The Canadian Engineer,* and *Canadian Machinery and Metalworking*.

These journals provide the greatest and most valuable source of information to the teacher and student. To illustrate this, a brief glance through the bound copies of *Canadian Transportation* in the 1920s gives the reader an immediate photographic impression of the development of snowplowing in Canada.

Other periodicals provide useful secondary sources of information. *The Canadian Historical Review* and *Ontario History* are examples of journals that frequently contain references if not whole articles that include technology. More obvious sources are found in publications such as *Canadian Rail* and *Canadian Aviation Historical Review*.

General

Some general sources are necessary to enable a student to relate Canadian events to world events. Therefore Ferguson's *Bibliography of the History of Technology* and Kranzberg and Pursell's *Technology in Western Civilization* are essential and reliable although both are shockingly short of Canadian content.

Agriculture

The agricultural history of Ontario is well served
by Jones and Reaman. One rare source refers to Quebec,
i.e. Evans, and I know of none that adequately cover the
maritime provinces. Grant MacEwen has contributed two
books which cover wheat production and tractors in western
Canada for a general audience and are useful for introductory purposes. Therefore, the subject of agricultural
history in Canadian provinces, with the exception of
Ontario, has barely been touched. Not even a survey of
the farm machinery industry is available.

Lumbering

The lumbering industry in Canada has received more
attention. Arthur Lower and Charlotte Whitton cover the
early square timber and sawn lumber phases in eastern
Canada but little secondary material is available for the
pulp and paper industry. Taylor's history of the industry
in B.C. has been long awaited but it is difficult to
evaluate as it contains no footnotes or bibliography. Much
more research is needed on the development and manufacture
of sawmill machinery.

Mining

The annotated bibliography on the History of Mining
in Canada that I completed in 1974 was an attempt to encourage further study in this area but I would request
others to comment on its usefulness. The history of the
Geological Survey of Canada is admirably chronicled by
M. Zaslow in a monumental work.

Chemical and Metallurgical Engineering

The distinctions between Chemical Engineering and
Metallurgical Engineering are blurred in Canadian history
but it is correct to say that both have been closely related to the mining industry. A *History of Chemistry in
Canada* by Warrington and Nichols illustrates that a large
proportion of applied chemistry in Canada is directly connected to our mineral production. Most of our mining histories stop short of the smelting and refining processes
and a good history of these important industries is badly

needed.

Building

The history of building and general construction is poorly recorded. Remple's book is excellent but covers only wooden construction in the early years. *Canada Builds* by Ritchie is a very broad survey which could serve as a starting point for several histories. I have no good sources for construction in western and northern Canada.

Surveying

Surveying in Canada has the best historical reporters of all. Thompson's three volumes are very well written and useful. The early reports of the Ontario Land Surveyors are interesting because it would appear that the members were often county engineers and their discussions on road and bridge construction are very useful.

Transportation

A large segment of the history of technology in Canada can be grouped under the general heading Transportation. Few countries in the world have had such obstacles to overcome and thus this industry alone can contribute much material for study. Glazebrooke's study is an obvious starting place and serves as a useful background text. However, it is forty years out of date and is very weak on air transportation, an area where Canada made a significant contribution to the world. There have been several government reports and Royal Commissions on transportation that could be considered.

Roads

Guillet's book *The Story of Canadian Roads* is a useful secondary source even if it is very general and it can be supplemented with material from several periodicals. It contains a very good bibliography. Two areas that are practically untouched are bridge construction and wheeled vehicles and in both of these areas it is incorrect to assume that they are direct copies of American technology.

Canals

The history of canal construction in Canada has not

been thoroughly covered. Some brief histories of individual waterways have been published and Heisler's *Canals
of Canada* is a reasonably complete overview. No thorough
study of the St. Lawrence Seaway is available for our
purposes. Many of the works do not give details of construction methods or effects on related technologies. Almost none refer to the use of the canals as a source of
waterpower and thereby analyze their technical effect on
local industry.

Shipbuilding

There are now several books available on sailing ships
or paddlewheelers that deal with areas of Canada from the
east coast to the Rockies. Very few actually discuss the
technology of shipbuilding or the details of their power
plants. There is no good account of shipbuilding in the
west coast. The sources listed are the most useful.

Railways

The history of railways in Canada has attracted
scholars, rail buffs, TV writers, and others whose material
covers much of the economic, social, and technical histories of railways and rolling stock. One area that has
been neglected is roadbed construction. The details of
construction, ballasting, trestles, bridges, and track-
laying are not easy to find.

Bradwin's story of the life of a railway worker is
one of the few of its kind in any field of technology. You
may be interested to know that Dorman's very useful book
is being updated by the Canadian Institute for Guided
Ground Transport here at Queen's.

Aircraft

Voyageurs of the Air by Main covers such a broad area
that it serves me as an encyclopedia of flying and flying
services in Canada; it suffers from lack of detail but
would be a good background text. It is also not very analytical. There is abundant material in almost all areas
of this subject.

Secondary Industry

This field is almost completely untouched except for the odd company history. While Canada has not had a highly developed domestic secondary industry as found in Britain or the U.S., there were numerous factories producing farm machinery, steam engines, and saw mill machinery for over a century. Research is hampered by the lack of company records; little has been attempted since Donald's classic *The Canadian Iron and Steel Industry* of 1915.

Biographical Material

There is no shortage of biographical material on Canadian engineers, inventors, and technologists. Most of the sources used in this paper plus the files of the Dictionary of Canadian Biography contain relevant material. In addition to technical detail much can be learned about Canadian attitudes to technology, the promotion of technology, and the role of technology and technologists in society from biographical studies.

Visual Material

By far the most useful collection of photographs is held in federal, provincial and university archives--useful because they are catalogued and labelled. Too many old photos have no date or description. Some private collections such as that of the C.P.R. are useful and the staff are usually very cooperative. All of these can be supplemented by photos and slides taken by the teacher from illustrations in periodicals and from existing artifacts and sites.

The National Film Board has several very useful films such as "The Forest," "A Salute to Flight," and "The Gold Seekers." Often only a portion of the film is factually informative but it may be very valuable by giving the student a feeling for the atmosphere of a period in history.

Maps are also very useful for teaching and for research and are usually available in archives. More general coverage is found in historical atlases.

Artifacts and Historical Sites

The historian of Canadian technology is very fortunate to have a wealth of material in museums and field sites. Although field trips are expensive in time, they are well worth the effort. Almost every area has a canal, railroad station, lime kiln, highway bridge, pioneer village, blacksmith forge, or mill site available to visit. If the students are thoroughly prepared, neither their time nor the teacher's time is wasted.

Another source of material that is often ignored is the visiting lecturer. There are literally thousands of people available with expertise and illustrative material that can be woven into a course. I have found the staff of the National Museum of Science and Technology especially helpful.

Conclusion

In summary, while there is more than enough material for several courses in the history of technology in Canada, there have been very few teachers and historians analyzing the macro structure of technologies in relation to overall Canadian history. One obvious void is in the relationship of one technology to another, e.g., the effect of the lumber industry on agriculture or the effects of the mining industry on transportation or on secondary industries, or the effect of road construction on agriculture, and so on. The relationships between the various technologies and the social, economic, cultural, and political life of Canada have also not been sufficiently explored.

There is another area that would be fascinating to explore in depth. What were early Canadian attitudes toward technology? Many of the nineteenth century British attitudes glorifying technology must have been shared by Canadians. Did our political and business leaders consciously perceive technology as their chief weapon to subdue the wilderness and expand their influence? The flamboyant editorials in the technical journals would provide one insight into the question.

All these relationships are a vital part of any good

course but no teacher can do all the research necessary. There is a desperate need for a vehicle of communication between teachers. This communication must also carry reviews of all materials, written and visual, as they are published. I believe the HSTC Bulletin has made a brave start and needs our support.

COMMENTARY:

P. J. Bowler

The concern for undergraduate teaching is important: if the history of science and technology is to be used as a means of bridging the gap between the "two cultures," it cannot remain as an esoteric field accessible only to a few graduate students. But this generates problems, since the more junior students do not always have the maturity needed to cope with the unfamiliar topics. Science and engineering students can be given standard works in Canadian economic history to fill in the background against which they can study the history of science. Arts students, on the other hand, are frequently terrified of encountering anything connected with science and thus require careful handling. Richardson is probably correct in asserting that they can, in fact, pick up an elementary understanding of the technical matters as they go along, but overcoming their initial apprehensions can sometimes be a more difficult task than he implies. In fact, given a free choice, arts students will generally shy away from a course in the history of science or technology precisely because of this fear.

Under these circumstances, the task of the teacher is indeed a difficult one. He requires not only detailed information that he can synthesize into his own lectures, but also suitable reading material to which the students can be exposed as they familiarize themselves with the field. As Richardson points out, suitable literature of this kind is not generally available. A great deal of detailed primary and secondary material can be found once one begins to look for it, and the bibliography given here is a valuable guide. But most of the secondary literature

is highly specialized, seldom written by professional
historians familiar with the issues of fundamental importance, and even more rarely written with the aim of intro-
ducing such issues to the new reader. We need more interpretive, analytical material even at the specialized level,
to give teachers a chance to put together meaningful
courses in the field. But beyond this, we need material
which relates the issues at a level comprehensible to the
introductory student at the university or even in the
high school.

These points are perhaps even more crucial in the
area of the history of Canadian science, to which Richardson
did not address himself. Here there is a host of important
issues which will certainly repay further study: government
and science, scientific education and organization, as
well as Canadian contributions to the various areas of
science. It is not quite true, as Richardson implies,
that nothing is available to cover these areas. A careful search will reveal a certain amount of material, although probably less than in the history of technology.
But again, most of what has been written is of only limited
value to the teacher and is virtually worthless to the
introductory student, because it provides no analytical
framework within which to understand the detailed events
reported. There are, of course, notable exceptions among
the more recent literature, especially Morris Zaslow's
Reading the Rocks. The collection published as *A Curious
Field-Book* is also valuable to give students an easy introduction to a carefully selected collection of primary
sources. But clearly, what is really needed is a comprehensive package of material providing analytical studies
of key issues in the history of Canadian science and technology. Those of us who have some specialized knowledge
in particular areas must be prepared to share our expertise
with teachers trying to put together a course in the field
for the first time. We must also try to provide reading
material that can be passed on to the students with some
degree of confidence that it will enlighten rather than
confuse them. It is to be hoped that the discussions at

this conference will lead to an active policy aimed at the production of such material. Without some such positive effort, the field will continue to stagnate in its present only partially active state.

DISCUSSION

According to the rapporteur, Vittorio de Vecchi, the workshop voiced two areas of concern. The first is that the materials necessary for research (e.g., business records, papers of societies and institutions) need to be continuously collected. One form of documents useful for both research and teaching is the biographical memoir. A suggestion was made to encourage individual scientists and engineers to write autobiographical accounts which could be centrally stored and catalogued. The collection would be announced by means of an analogue of *Dissertations Abstracts* and the typescripts could be made available to researchers and teachers on microform. A similar proposal called for the Royal Society of Canada to maintain current biographical files similar to those of the Royal Society of London.

A second area of discussion was the need for readily available classroom materials. After recording its pessimism with regard to a textbook for the field, the group turned to a general plan for producing short, flexible, and cheap modules along the lines of the British SISCON series. This was heartily endorsed, though the consensus was that these modules should be, at least initially, strictly limited to history of Canadian science and technology rather than broader issues of science and society.

The history of Canadian science and technology is approximately at the stage of the history of American science in the late 1950s. Our American colleagues faced many of the problems we do: the professional historians had taken scant notice of their field and some were openly scornful of the idea of an identifiable "American Science." The early works in the field were relatively unsophisticated (e.g., Dirk Struik's *Yankee Science in the Making*), since a great amount of historical detail and theoretical focus

had not been developed. In twenty years' time, that field
has come of age. We, in Canada, cannot afford to wait
twenty years, so we must learn lessons from them where
we can, and improve upon their approach.

Since no one yet feels sufficiently capable of undertaking a book even of the level of Struik's, might we
accelerate our programme with the module scheme? The
Americans had nothing like it. However, discussion at the
conference on which topics might be cultivated reminded
one of Stephen Leacock's horseman--riding off in all
directions. Instead of narrowing our view, as was suggested, we should do just the opposite: we need works on
theory, works full of facts, and works on the social aspects of science and technology. While such a mélange of
modules may lack uniformity, they would stimulate discussion and research more effectively than a large-scale
textbook or, what would amount to the same thing, a series
of modules along similar lines.

COURSES IN THE HISTORY OF CANADIAN SCIENCE AND TECHNOLOGY:
THEIR PURPOSE AND CONTENT

Richard A. Jarrell

Most of those attending a conference such as this
would make the teaching of Canadian science and technology
in the university a high priority. Many academics may
be willing and able to initiate such courses, but those
who have already accomplished this realize that the
establishment of such courses is no easy matter. Before
taking up the more practical aspects of this problem, I
think a few words on the politics of curriculum planning
are in order. Before anyone undertakes to plan and mount
a course dealing with Canadian science and technology, he
or she must be clear about the importance of the subject.
A simple appeal to try something new will make no impression upon curriculum committees in these lean times. The
first advice must be to think out a rationale carefully.
Some useful ammunition will be found in T. H. B. Symons'
report *To Know Ourselves*, but one's arguments must be
tailor-made for the particular context. In Canadian universities, these contexts can be broadly classified under
three headings: science or engineering programmes, history programmes, and interdisciplinary programmes. Each
has its inherent obstacles for the would-be innovator.

To attempt to mount a course on Canadian science and
technology within a science or engineering (or even medicine) programme is perhaps the least likely path to success, since scientists and engineers seldom perceive much
connection between their notion of what a science or engineering student should learn and the social scientist's
notion. Adding Canadian content to existing courses on
science, or on the history of science and technology, or
on science, technology, and society offers fewer

complications. Few independent courses are likely to
arise in this context, but some Canadian content in
existing courses would be a distinct improvement. Internal departmental opposition to this minimal curricular
change should be slight, but the planning burden on the
teacher is still large considering that the integration
of Canadian material into a traditional course may be
difficult.

The obvious context for courses in this field is history, but few courses have arisen in history programmes;
those most interested in the subject of Canadian science
and technology are not often members of the history department and historians often display some hostility (or
at least puzzlement) about the field. In Canada, political
and economic history reign supreme, and innovations are
rare. If faced with vested interests of this kind, the
instructor may have to compromise by adding scientific
and technological material to existing formats. This
would be welcome enrichment, particularly for social and
intellectual history of Canada courses. If a full course
on the subject is insisted upon, then the faculty member
will have to legitimize the subject. Unfortunately, the
student response to existing courses has so far been dis-
couraging, offering little support for the introduction of
new courses. Like it or not, salesmanship may have to be
placed before academic justification.

The third context, that of interdisciplinary studies,
is possibly the easiest route to the calendar, but it has
its own peculiar problems. Programmes in Canadian Studies
or in science and society provide the usual examples of
this context. There is a sense in which university committees welcome curricular oddities into such programmes, but
the financial and enrolment bases for such programmes is
all too often precarious. These programmes often reflect
intellectual needs of the sixties rather than the seventies,
and sometimes reflect American rather than Canadian approaches to subject matter. No matter how sound some of
us feel these programmes to be intellectually, they will
always be considered marginal--and in a pinch, expendable--

by our more conservative colleagues. To compound this
problem, few students perceive their educational goals in
the manner in which they might have a decade ago. What
does one say to the student who asks what practical benefit
he can derive from a course on Canadian science and technology? Nonetheless, such programmes may be the last re-
sort for some instructors.

The more practical matter of course planning, the
length of the course, the topics to be covered, all depend
upon the context, the probability of success in attracting
sufficient numbers of students, and the teacher's competence in the field. There is a wide variety of formats
that might be adopted, but they fall into six major categories:

1. Courses on Canadian science--historical
2. Courses on Canadian science--contemporary issues
3. Courses on Canadian technology--historical
4. Courses on Canadian technology--contemporary
     issues
5. Some combination of any of the above
6. Some combination with the addition of medical
     material

These categories cover virtually all the courses currently
offered in Canada. Unless the competence of the instructor is particularly great, too many topics will result in
superficiality. At this point in the evolution of the
field, none of us can claim competence over the full range
of subjects enumerated above, but it might be argued that
even a superficial presentation of many topics is better
than nothing. This is a personal decision.

Despite the subject matter ostensibly covered in a
course, there can be courses with little content. The
high-content course, which emphasizes reading, lectures,
and discussions, is typical--and more suitable--for the
undergraduate. The low-content course, which emphasizes
individual research with the instructor often taking a
passive role, is appropriate on the graduate level. I do
not wish to deal here with graduate courses but there is
one problem with low-content courses. The student of such
a course may have a good science background but be woefully
ignorant about Canada, or *vice versa*. In either case, the

final research product is bound to be inadequate. There
are remedies: one might require a brief (two months or
so) intensive reading programme before beginning the
seminar format or, alternatively, the instructor might
require prerequisites in science, Canadian history, or
engineering. The field, if it is to mature, requires good
graduate students and high-quality research from them, not
half-digested material that does no one any good. Other
graduate courses have prerequisites and those on Canadian
science and technology do as well. We will all agree, I
am sure, that standing at the frontier as we are, our
standards must be high.

Undergraduates will come into an undergraduate-level
course with even less preparation, so not quite as much
can be expected of them. Many upper-level courses in history, for example, are run on a seminar basis. For our
subject, this format would be a mistake for the same reasons that apply on the graduate level. Undergraduate seminars may be useful if there is a lower-level course available on Canadian science and technology, but few of us can
expect to be so lucky in obtaining two places in the curriculum. Thus, the high-content course ought to be the
norm.

Having taught several courses of this kind myself,
and having studied the efforts of others, I suggest that
the third year is the optimum place in the curriculum.
The students will be more mature, have research and writing
experience, and have had some exposure to science, or history, or engineering. From my experience, students at this
level maintain a high level of enthusiasm and the essays
they produce are of high quality; indeed, I have seen some
just short of being publishable. If a course is offered
to first or second year students, many of the goals of a
course will not be met. It is difficult to maintain enthusiasm for something you know nothing about. Another tacti-
cal error would be to allow such a course to be a substitute for a science course for arts students, or as an arts
course for science students. It might seem expedient to
some, but it can only compound the ignorance of the

undergraduate who comes out of secondary school blissfully
unknowledgeable in most things. The fortunate instructor
will be he who has students who are functionally bilingual
because the literature in the field is almost entirely
English-language discussions on science and technology
in English Canada, and French-language discussions of
Quebec science and technology.

The course content need not be entirely historical;
history might be a minor theme, or altogether missing, but
the caveats enumerated above will still apply. These
problems take on different dimensions when one is adding
Canadian content to a more general course. In some re-
spects, the Canadian-content course takes more careful
planning than a full course. Regardless of whether the
course is historical or not, and whether it deals entirely
with science and technology in Canada or not, there exist
several distinctive approaches to the subject matter, four
of which are briefly described here:

1. *Chronological Approach:* This is the commonest and
most straightforward manner of presentation. There is a
danger of trivializing the subject matter by attempting to
cover too much in too short a time. Of course, the instructor must be creative to turn a chronology into history.

2. *Topical Approach:* This can be chronological in
structure, but one typically eschews the broad sweep of
events for a few representative topics. The choice of
topics may depend upon the resources available, the instructor's knowledge, or his own predilections. The weakness in this approach is the spottiness of the literature;
there is a danger of giving a false impression of Canadian
science and technology as a whole.

3. *Time-Slice Approach:* Not often employed for pedagogical purposes, this method provides historical "snap-
shots" at several (presumably) pivotal stages of the growth
of Canadian science and technology. The device is more
literary, a recent example being Robin Harris' *History of
Higher Education.* In many ways this approach has much to
recommend it, but the gaps it leaves require a great deal
of creative history on the part of student and teacher

alike.

4. *Local Interest Approach*: This approach concentrates upon science and technology in the university's
locale. This may be largely institutional history, or
provincial, or regional. Depending upon the locale, the
resources for such a course may be extensive but one loses
a great deal of perspective. This approach may be most
valuable in a Canadian-content course rather than the
basis for an entire course.

The weaknesses of any one approach can be offset by
combining two or more of them in one course. While re-
quiring more planning, the multiple-approach course should
be richer and more rewarding to students willing to work
hard. Whichever approach is adopted, there is still the
historiographic question. Broadly speaking, there are
two ways of looking at the history of science or technology (and at contemporary issues, as well): the so-
called internalist and externalist views. Simply put, the
former follows the development of specific ideas or techniques, while the latter attempts to explain the social
and intellectual context of discovery and development.
Unless our view of Canadian science, for example, is completely in error, the internalist approach would yield
little and would bore the student. Yet, there are a few
aspects of Canadian science, and many topics in technology,
that would benefit from an internalist analysis. The
externalist historiography will be far more valuable in
university courses, but some kind of balance should be
struck. Unfortunately, the literature is too thin at
present to allow a good balance between the two views.

Finally, I wish to turn to the practical matters of
course organization. My outline assumes that the course
is devoted entirely to Canadian science and technology,
both historical and current, at the upper level. Given
the usual unpreparedness of students, lecturing is indispensable. The instructor must impart both factual material
and a framework for the facts. Admittedly, this is not
easy due to the paucity of readily available literature,
but lectures cannot be replaced by tutorials or seminars at

the undergraduate level. Tutorials are essential, however, and may range from those with little structure, in which students chat about the readings, to the highly structured, in which students report on preassigned material. An instructor may find that it is necessary to loosen up or crack down in midstream depending upon the students in the course. My own experience is that the highly structured tutorial works best for students with sketchy backgrounds; it prods them into research beyond the texts assigned and allows them to pursue in more detail topics that arise in lecture at a superficial level. Some teachers are highly successful at drawing out student discussion of high calibre without resorting to such a mechanical arrangement. Most may find a compromise solution the best.

Individual research is necessary at the undergraduate level. Beyond the conventional wisdom that research and writing are good for students, there is a bonus for the instructor in a young field such as ours. New bibliographic sources will be unearthed and many insights will be gained by both students and teachers. In large courses, it is often the case that essay topics are assigned, or a small selection of topics is offered for student choice. This would be counterproductive in courses on Canadian science and technology. These courses tend to be small, student background may be highly varied, and the instructor's knowledge will have many lacunae. Allowing open season on essay topics is bound to result in a richer educational experience for all involved.

There are a number of teaching aids besides library resources. Films and slides for our field are virtually nonexistent, though the photographically-inclined instructor could make up sets of slides of scientific or technical items of interest, or have the university audiovisual department make up slide sets from published photographs. Printed charts or overhead transparencies can be prepared for graphic or statistical information; these would be particularly useful in discussing economic or science policy issues. Courses specializing in technology have potentially more resources than those in science since there

are always examples of Canadian technology at hand in
museums, pioneer villages, forts, historic sites, factories, dockyards, railroads, bridges, architecturally
interesting buildings, etc. While these are more concentrated in central Canada, every university locale in the
country has a wide variety of examples nearby. Science is
not so well endowed, though museums are often useful.
There are archives scattered throughout the country and
nearly all of them have some holdings in science and
technology. The use of archival materials by undergraduates enhances their educations considerably.

The foregoing is by no means all that can be said on
the subject of courses on Canadian science and technology.
Our greatest desideratum is, however, a textbook. Our
field is not emerging in a way comparable to some other
interdisciplinary fields such as ecology, for such fields
already had extensive primary and secondary literatures
before they caught the wider imaginations of university
audiences. Many of us believe that a knowledge of the
nature of science and technology in this country is critically important. We cannot afford to wait for the field
to evolve over a period of time similar to that of the history of American science and technology. The sooner we
can interest the undergraduate, the sooner we will get
graduate students willing to study Canadian topics. Thus,
we have a dilemma to overcome: courses on our subject
are difficult to teach with a weak literature, but a
strong literature will not be forthcoming until we provide
outstanding courses.

COMMENTARY:

Philip C. Enros

I certainly believe that my inexperience explains the
main impression I have of this paper--that it is concerned
with *very* practical matters. Indeed, I think that that is
the paper's chief fault. For it is overly concerned with
course tactics to the extent that it completely neglects
a discussion of the purpose or goals that courses in the
history of Canadian science and technology ought to

pursue. In other words, I think that it is important to
be able to respond to the question that Jarrell's student
posed: What practical benefit can one derive from a course
on Canadian science and technology? The instructor should
have a response to this question not simply for the sake
of promoting these courses or of justifying their existence
to curriculum planners but because he has given some
thought to the meaning of the course, to its purpose, within the context of the student's university education. Now,
apart from such practical aspects as obtaining research and
writing skills, it seems to me that the fundamental purpose
of any course in the history of Canadian science and technology must be to give the student an understanding of what
is meant by the terms *Canadian* science and *Canadian* technology. This purpose, I believe, may be achieved in two
ways. First, by relating the science of persons in Canada
as well as their image of science to Canadian religious,
social, political, and economic thought. This relationship may be explored through such avenues as the study of
the reception of scientific ideas or of the role of such
institutions as government agencies, scientific societies,
or universities. Out of this interchange will come the
concepts and themes that will define and characterize
Canadian science and its development. Secondly, our goal
or purpose will be attained by contrasting the Canadian
scene with other national styles of science. And here,
within an internal perspective, I am quite sure that while
Canadian technical achievements or concepts might not
always be very significant from the point of view of the
overall development of a particular science or technology,
still, they would hardly be boring. For they would reveal
the ways in which the Canadian scene determined Canadian
science and technology. Thus, my chief comment is that
while courses in the history of Canadian science and technology may be in their infancy, what they really require
for their proper development now is a clear picture of the
purposes which they should serve.

Finally, as a sort of footnote, I would like to pick
up Jarrell's plea for a textbook. The lack of a

satisfactory textbook plagues even the older discipline
of history of science and technology. A text would be
of great benefit for the development of Canadian courses.
I would like to suggest that a text might be well and
easily developed as a series of units, for example, on
the model of the Open University courses or the SISCON
texts. Each unit could be prepared by a different author
or team of authors, thus allowing for the combination of
specializations and the quick production of a quality text.
Jarrell's work, his paper today, and his forthcoming bibliography all illustrate the extent to which the elements
for such a text already exist. Therefore while past experience may support a pessimistic view, there is much to
hope for and much promise for the future of courses in the
history of Canadian science and technology.

DISCUSSION

The rapporteur, Ian MacPherson, reported that some
members of the workshop believed that the introduction of
courses may not be as difficult as suggested; others pointed out that the current diversity in offerings might be
preferable to a more unified approach. The participants
called for more professionally oriented courses to help
produce workers for specialized fields such as museum
work and historical reconstruction. An intense exchange
of ideas regarding the differences in potential audiences
ensued, particularly noting the differences between arts
students on the one hand, and engineering and science students on the other.

Interest was also generated on the possibility of
short courses or chautauquas for non-specialists. These
could be held in the summer and might be staffed by
Toronto and Montreal area specialists.

A problem with this workshop was the total lack of
discussion about education in a wider sense. The participants disagreed on the degree of optimism one might bring
to the course-planning exercise, but even in the best of
all possible worlds, the total number of students we can
ever reach will be small. If there is any social utility

to this field--and we emphatically believe there is--then
we must reach far larger audiences; university courses are
not the proper vehicles for this. We must employ all media
available to us. Radio and television (and increasingly
cable) provide outlets for both scientific and historical
material. If CBC and Radio-Canada can deal with science
and technology in contemporary Canada, then they might be
persuaded to look into the past. Certainly the technical
triumphs and tragedies of the transcontinental railroad
are every bit as exciting as the political machinations.
Well-written articles for the popular press, of which
Canadians are avid readers, would be an important outlet.
Public lecturing to local groups instead of to bored historians at conferences is an experience we should all
have. Canada is the most "wired" country in the world;
our potential classrooms are vast. Let us enter them.

A.6   ARCHIVES

PROBLEMS FACED BY ARCHIVISTS DEALING WITH
HISTORICAL RESOURCES OF SCIENCE AND TECHNOLOGY IN CANADA

Sandra Guillaume

I would have much preferred that the title of this
effort should have been "Some problems faced by archivists
dealing with the historical resources of science and technology in Canada," because I am certain that the problems
as I see them are only *some* of those faced by archivists
in dealing with the historical sources of science and technology in Canada.

If one accepts that the logical places one might expect to find source materials for the study of the history
of science and technology are the archives, libraries, and
museums of the country, a brief look at the histories of
these institutions may give an historical context to this
examination. Lewis Thomas' survey of archival legislation
in Canada noted that the first legislation relating to the
disposition of government documents, i.e., archives, was
passed in 1790, and, although Nova Scotia established its
archives in 1837, to be followed by the Public Archives
of Canada in 1872, and various provinces in turn, the
archives of Alberta and New Brunswick were Confederation
centennial projects, while Prince Edward Island's was established for the centennial of the Charlottetown conference
in 1964. There may be some dispute about this, but the
first university archives charged primarily with the collection of university records and the papers of professors and
staff was set up at McGill in 1962.

Although neither librarian nor museologist, I would
note that the National Library has but recently been cele-
brating its twenty-fifth anniversary, while the National
Museums of Canada have only within the last ten years or so
been truly active in other than ethnological, archaeological,
and similar activities. The Ontario Science Centre was

177 .

another centennial project. Finally, if I may be permitted a personal note, I can vividly recall the hair-
tearing efforts of the members of the Council of Campus
Archivists at the University of Toronto in 1976 to frame
a resolution urging the University to establish a policy
regarding the preservation and collection of scientific
devices, teaching aids, lab equipment, etc., still on the
campus.

It may be seen from this sketch that the nonexistence
of agencies specifically charged with the responsibility
of seeing that papers, printed materials, and three-
dimensional objects are collected, preserved, and in some
way made available for those who wish to study them hampers
researchers generally but more particularly and acutely
those in scientific and technological areas.

Why more acute? I venture to suggest that there is
a real and two-headed educational problem; namely, the
education of the archivist and the education of the
scientist. Fairly few archivists of my acquaintance have
any university-level science or history of science, and
the great majority of the scientists I have met really have
not had any good grounding in history or history of
science. The archivists are somewhat hesitant about
attempting to acquire scientific and technological material,
and scientists have tended to file everything in the 'round
file' (wastepaper basket to the uninitiated) or cannibal-
ize equipment, leaving the remaining parts to the tender
mercies of janitors, and to basements and attics of less
than optimum climatic conditions.

Three-dimensional objects also are not normally part
of the holdings of an archives and although the archives
in which I have worked have all taken this type of material, it is perhaps best housed elsewhere. Or should it be?
Is it better to house the scientist's instruments with his
papers? And what about slides, specimens, etc.? The end
result of all this is, of course, paucity of sources.

Generally speaking material which finds its way into
an archives was not created specifically to be put in the
archives but rather for some immediate transaction. This

is as true of a scientist's experimental notebook as
the account books of a commercial firm. In both cases, the
creating office considers that it has some right to hang
onto the material for a period of time. In fact, it
probably has to keep the record to hand for some time.
This can result in the pack rat syndrome amongst records
creators. It most assuredly has resulted in a salvage
syndrome amongst archivists. In the former case, pack
rats hang onto materials, frequently in quarters designed
for other purposes, and the end result is a salvage operation frequently complicated when a fire or some other
catastrophe occurs. In this connection, fires such as
those at University College, University of Toronto in 1890,
McGill University's Faculties of Medicine and Engineering
in 1911, and the Parliamentary Library in 1916 may be
written off by some as due to outdated buildings and lack
of firefighting equipment. Much more recently, however,
in 1977, the Sandford Fleming building went up in smoke,
resulting in as yet undetermined losses of papers of
engineering significance. I am not suggesting that scientific faculties in universities deliberately self-immolate,
but the phenomenon perhaps deserves further study and
obviously cuts down on materials available for research.

   Most archives today in one way or another are involved
in some form of records management, that is, the planned
disposition of records. This is true of government,
business, and university archives, and involves the systematic weeding out of those records which can be destroyed
after a set period of time when their utility is over and
transferring to an archives those of permanent value.
While I am quite convinced of the effectiveness of this
planned disposition respecting those administrative records
there is a question of whether this works for "personal
papers," and what and whose are the "personal papers"?
This is a question which has vexed many archivists,
scholars, politicians, etc. It can scarcely be denied
that the papers of individuals are an important element in
scientific history, as well as in other branches of history
The problems of systematizing the acquisition of personal

papers, i.e., papers of professors, scientific officers,
etc., are perhaps greater than those of administrative
records, but surely can be overcome. Along with the
problems of paper acquisition, some individuals do not
leave papers as such, yet can contribute through the oral
history interview fascinating and important details. I
think particularly of the sidelights illuminated by inter-
views of the engineer who worked with Dr. Norman Bethune
to develop instruments for his experimental surgery. Many
persons are still alive who were actually involved in
significant scientific and technological advances and
could be tapped, were funds and skilled interviewers
available. How best to tackle this? A team approach with
one questioner a scientist in the particular field, the
other a more general interviewer?

In answering a questionnaire recently, a colleague of
mine referred to archivists as being pragmatic. Very few
archives ever consider that the force of their legislation, budgets, and staffing permit them to attack all
areas of responsibility at once. To a degree, then, the
use of materials and the patterns of research will likely
determine to some extent the acquisition pattern of an
institution, and also the priorities of processing. The
fact that interest in the history of Canadian science and
technology is recent is reflected to some degree in these
patterns.

Granted that the collection and preservation of materials must come first, there still remain the problems of
arrangement and description. Having in mind the educational problem earlier referred to, there may be some difficulties in adequate descriptions being arrived at. With
mutual interest and cooperation between archivist and
scientist, this, as many of the other problems mentioned
here, can be jointly attacked and hopefully solved.

Hopefully too, this brief summary will provide a focus for discussion and stimulate the formulation of defini-
tive resolutions of some of the problems raised.

COMMENTAIRE:

J. Bernier

Sandra Guillaume a soulevé dans son exposé des problèmes importants. Les législations canadiennes et provinciales relatives à l'acquisition et à la conservation des archives en histoire des sciences et de la technologie sont jeunes, généralement vagues et souvent inefficaces.

Le fait qu'on ait demandé à un historien de faire le commentaire de cette conférence confirme aussi le problème du manque d'archivistes spécialisés en ce domaine dans plusieurs parties du pays. Dans certaines provinces en effet, dont le Québec, il n'y a même pas encore d'archiviste dont la fonction soit de voir à l'acquisition, à la conservation et à la classification de ce type d'archives. Je dirai même que, dans certains dépôts d'archives provinciaux, ce sujet ne figure même pas au catalogue.

Jusqu'ici, ce sont souvent des institutions privées (ou jadis privées) qui ont vu à la conservation de ce type de documents. Ce sont, entre autres, les universités, les hôpitaux, les associations professionnelles, les centres de recherche et certaines entreprises. Mais il n'y a pas de politique globale pour les régir et les archivistes désignés, malgré leur bonne volonté, n'ont pas toujours le temps et la formation nécessaire pour bien remplir cette fonction et répondre aux diverses demandes des chercheurs.

Les malaises qui prévalent, par ailleurs, dans les dépôts provinciaux et fédéraux ne sont cependant pas attribuables aux seuls archivistes et responsables de ce domaine. Les historiens, il faut le dire, ne les ont guère stimulés et aidés dans leurs tâches. Jusqu'ici en effet, les historiens canadiens de l'histoire des sciences et des techniques ont surtout travaillé à partir de sources imprimées, de sorte que le besoin ne s'était pas tellement fait sentir de voir à l'acquisition de telles archives, à leur classification et à l'élaboration d'instruments de travail pour faciliter leur utilisation. Par contre, je suis assuré que des progrès importants pourraient être réalisés dans le développement et l'organisation de ces

archives si les historiens les utilisaient d'une manière
plus systématique, et s'ils faisaient davantage part de
leurs besoins et de leurs suggestions aux archivistes.

Il importe aussi d'entreprendre une action auprès de
la population, des entreprises et des institutions concer-
nées afin de les informer de l'existence de tels dépôts et
de les inciter à venir y déposer leurs archives. Ainsi
confiés à des centres spécialisés, ces documents seraient
davantage accessibles aux chercheurs.

En somme, même si la situation des archives s'est
considérablement améliorée au cours des dernières années,
le domaine des sciences et de la technologie fait encore
figure d'enfant pauvre. Cependant je pense que, grâce à
des contacts plus étroits entre les archivistes, les historiens, les scientifiques et la population, nous pourrons
avancer rapidement et rattraper peut-être le temps perdu.

DISCUSSION

A good deal of the discussion, according to David
Rudkin, centred on questions dealing with the *raison d'être*
of an archive. According to one school of thought, their
primary function is to serve their sponsoring institutions,
not historians. Where archives have neglected scientific
and technical material, it can be as much the fault of
historians as the archivists since, until recently, they
were uninterested in the problems of this field. The situation is now exacerbated by the general lack of sufficient
staff, both at the Public Archives of Canada and at the
many other archives in Canada that could potentially serve
the historian of science and technology. There was disa-
greement on whether the real problem was a lack of money
or a lack of commitment, strength, and imagination.

The consensus of the workshop was to encourage individual, rather than amorphous collective, action to secure
better archival holdings in the many local archives. Furthermore, much could be done to stimulate the growth of
business archives.

The issue of the Robert Bell papers emerged during
the session and resolutions were sent to the plenary

session. There, the assembled conferees heard that the
Public Archives of Canada had once accepted the papers of
the noted nineteenth-century geologist Robert Bell, but
had subsequently given the bulk of the papers back to the
family. The papers were then offered for sale at a public
auction. The plenary session passed resolutions--which
were dispatched to the responsible government figures--
demanding that action be taken to recover these valuable
papers. (The Public Archives has subsequently authorized
selective bidding on the Bell Papers.)

THE PRESENT STATE OF MUSEUMS OF SCIENCE AND TECHNOLOGY
IN CANADA AND THEIR USE AS
RESOURCES FOR HISTORICAL STUDIES

C . J . B . L . Porter

Beyond the study of a few subject areas little is
being done by our museums at the present time to examine
the broad field of the history of science and technology
in Canada. Museum personnel would be the first to admit
that this was the case. I think we all are readily aware
that we have no Smithsonian Institution, Deutsches Museum,
and no (British) Science Museum, but perhaps it is more
disturbing to recognize that under present conditions there
is little prospect of an institution of their stature
developing in Canada. Museums of science and technology
are enthusiastically attended the world over, but it is
often the case that few can balance their popular exhibit
programmes with in-depth research and publication. A list
of the museums of science and technology in Canada prepared
by the Canadian Museums Association is included in Appendix
B.

To use the museums as a resource for historical
studies we should take time to examine the museum field and
familiarize ourselves with their strengths and weaknesses.
The enquiry is made more difficult, however, by the fact
that the institutions which are conducting research relevant to science and technology often are not museums of
science and technology, and furthermore many of the collections which would be useful in supporting research into
this field are not located in institutions specializing in
science and technology.

My introduction to this workshop will be for the most
part limited to institutions based in Ontario. However I
would like to express my gratitude to the staff of Parks
Canada, to the History Division of the National Museum of

Man, and to Barry Lord of the National Museums of Canada,
who have provided me with information outside the province.
In light of this somewhat limited approach to the topic
I am hoping that those of you who are familiar with museum
collections and research in the field of science and technology in other parts of the country will share that
knowledge with us. In particular, information about the
programmes at the British Columbia Miners Museum and its
counterparts in Nova Scotia at Stellarton and Springhill,
at the Western Development Museum in Saskatchewan, and
the Manitoba Agricultural Museum would be particularly useful.

It is evident from what follows that some liberties
have been taken with the title of the introduction. Museums other than museums of science and technology and
some institutions which would deny with some reason that
they are museums have been included. This has been necessary so as to include the research into science and technology which has been done by these other museological
agencies.

What should we be expecting from our museums? Without
being waylaid into a discussion about the minutiae of what
museums are, we can state that they ought to perform three
functions. They should collect and preserve artifacts
representative of their field of interest; they should perform research on the collection which has been assembled
and on the material culture background of which the collection is an intrinsic part; and lastly, they should make
both their research and their collections available to the
public in as stimulating and informative a way as possible.
In reality few museums perform all three functions equally:
one museum will invariably perform some parts better than
others, while some deliberately set goals which assign different priorities to each of these areas. It is important
to recognize the priorities or limitations of institutions
when we approach them for study purposes. The Ontario
Science Centre and the National Museum of Science and Technology make an interesting comparison.

The Director of the Ontario Science Centre has stated

publicly that the Centre is not in the business of amas-
sing historical collections, but rather of specializing
in the popular interpretation of current developments in
science and technology. This policy is evident throughout
the Centre. The burden of interpretation is placed on
working models and audio-visual programmes, rather than on
historical artifacts. In the presentation of these current
trends their exhibitions are unsurpassed in Canada. The
National Museum of Science and Technology is more orthodox
in its approach and sets out to perform the three general
functions outlined above. Of the three, the interpretation
of the collection for the general public has been assigned
top priority while representative collecting and research
have become secondary museum functions. For the researcher
into the history of science and technology a recognition
of these policy decisions is important, as it will recommend which agencies should be consulted and for what purposes.

The collection and display of artifacts representative
of technological development frequently cause considerable
problems for museums. Too often lack of display or storage
facilities have led to some industrial and technological
processes which are both nationally or regionally significant being inadequately represented. The size of the machinery and the dependence of some processes on particular
environmental or topographical conditions have precluded
their satisfactory transference to museum display. The
hardrock mining and lumbering industries, both of which
are pre-eminent in the development of Canada and which produced a great variety of technological innovations, remain
almost unrepresented. The use of models and other exhibit
devices cannot always act as an effective substitute for
the real thing.

There are other more fundamental problems, however.
Until recently the absence of any comprehensive industrial
or academic interest in systematically identifying nationally or regionally significant developments in science and
technology has made the problems confronting museums collecting in these areas more difficult. Too often museum

curators feel that they are working in a vacuum. They
must attempt to evaluate the significance of the artifacts at hand without complete documentation and without
the general body of knowledge which develops as several
generations of specialists work in a field. As many of
you are well aware, technological research is both time-
consuming and complex due to the lack of primary documentation in conventional sources. Furthermore the location
of primary documentation has been too often buried uniden-
tified among documents with social, economic, and political
classifications. Curators working in mature art, natural
science, and human history museums have been able to dedi-
cate a significant part of their time to research; as a
result over the years they and their academic peers have
developed a body of knowledge on which succeeding generations of specialists have been able to build. But in
Canada this has not been true of museum curators working
in the newer field of the history of science and technology, and so the lack of adequate academic support is
keenly felt.

There is another factor that should be given consideration. With the possible exception of science centres,
there seems to have been insufficient rapport between
museums and industry in the past and little attempt to
come to terms with the social implications of technological and scientific development. Whether this has arisen
from a lack of industrial concern with these implications
or whether museums have not recognized the need to conduct
an adequately vigorous industrial collecting campaign, the
result has been the widespread loss of machinery and documentation. Instead museums have become the recipients
of other collectors' collections and have allowed others
to determine what is important in the field and have not
actively aroused industries' interest with their contribution to our heritage. In the absence of a working relationship museums are not considered a logical repository
for industries' outmoded but significant artifacts, and
for the technological data which is essential for their
interpretation.

Canadian museums of science and technology have been recent developments. In fact the two largest in central Canada are the products of the 1960s, and even now major collections in the fields of telecommunications, transportation, and aeronautics have either no place for public viewing or are housed in inadequate and temporary facilities which might jeopardize their survival. With the birth of these museums coinciding with a strong public demand for display and popular information, the climate has not been politically conducive to balancing the promotion of interpretation with adequate research. In older institutions of course the reverse has often been the case, leading to the accusation that museums are irrelevant to the public at large. Perhaps in the future a more equitable division of resources in these areas will be achieved.

However, let us find out who is doing what and where. The sheets which accompany this introduction outline which institutions are currently doing research relevant to our field and which have significant collections to support this type of enquiry.

Parks Canada is working in a number of areas related to science and technology in Canada. Their research is geared to serving two ends: the first is to satisfy the very particular problems which arise at historical sites within the Parks Canada system of national parks. The second is more general and is designed to investigate regional histories, trade patterns, and industrial processes, information which is required to set the interpretation of particular sites in an appropriate historical setting. Past research material can usually be found printed but not published, in the Manuscript Report series. Copies of the titles and the manuscripts can be found in the Parks Canada headquarters in Hull. Some of this material is further refined and eventually appears as government publications published by the Department of Supply and Services. Current research material and particular site studies may however not reach the Manuscript Report series and they then remain at the regional offices located

in centres across the country. Consequently it is useful
to know what sites fall within what region. Of particular
interest to this group might be the work done for the
Forges St. Maurice at Trois-Rivières, the building research
being carried out at Quebec City, Louisburg, and the
Halifax citadel, and the mining research in the Yukon.
Clearly for all of us in this field a knowledge of the
work of Parks Canada is of considerable importance. However just because they are a civil service department, it
does not follow that their research data is always available on demand.

The National Museum of Science and Technology's programming has been oriented towards public participation.
An implicit consequence has been the low priority given
to systematic research either into the collection or the
general disciplines in which the particular collections
are a part. Despite the lack of comprehensive research
where systematic collecting of artifacts has been attempted, the student of the history of science and technology
will find useful material.

The National Aeronautical collection of the National
Museum of Science and Technology at Rockcliffe is interna-
tionally important, and pre-eminent in Canada. Furthermore the collection is catalogued and research has led to
the publication of two books. The major technological
developments in aviation in Canada are represented. The
collection and supporting documentation is of considerable
importance to the student of science and technology in
Canada. Other significant collections in the Museum of
Science and Technology include agricultural machinery,
telecommunications, printing, and railway equipment.

The present acquisition programme is leading to a
number of collecting problems. First, significant technological developments in Canada associated, for example,
with hard-rock mining, forestry, snow transportation, and
other areas are either not represented or insufficiently
represented. Second, the museum's collection has developed
from gifts from well-intentioned donors rather than from
an established programme which identifies areas of national

importance and directs the acquisition programme towards
those areas. The consequence of this may be duplication
with provincial and other agencies or inadequate representation of national themes. Nevertheless the collections
are important and the curators in charge will be only too
happy to help you with your enquiries.

The Science Centre in Toronto is an anomaly. As suggested above, it is not strictly speaking a museum and has
virtually no historical collection. However, in its
approach to current technological development, and in its
introduction of science and technology to a broad audience,
its exhibits are unequalled. In its displays there are
several artifacts which are unique in Canada and important
in the history of technology; examples include a functioning Jacquard Loom, an electron microscope built by the
University of Toronto, an early design of wood lathe, a
"slowpoke" nuclear reactor, Banting and Best's laboratory,
and a holograph laboratory. Where this institution might
be of particular use is not so much in its collections and
exhibits as in its lecturing facilities. The Science Centre could promote the study of the history of science and
technology by organizing lecture series. The Centre has
already sponsored programmes on esoteric subjects, and one
of these which is planned for the near future is to be on
inventors and inventions. The development of these lecture
series is one of the principal methods by which the Science
Centre can reach academic or special interest groups in the
community.

Currently the Science Centre is planning a mining
exhibition. The displays will describe hard-rock mining
in Ontario and will thereby fill a need for activating
public awareness of mining within the province. However,
the plans do not include a historical collection of mining
artifacts.

With these two substantial institutions already established in central Canada and heavily funded by two levels
of government, it is unlikely that there is room for another organization which would make up for the lack of
historical research into science and technology. Let us

hope that the proposed Museum of Industry and Technology at Stellarton, Nova Scotia and other institutions across the country will take up this neglected challenge.

The History Division, the research branch of the National Museum of Man, is conducting some research relevant to our discussions. Areas of study are detailed on the accompanying sheets but of particular interest might be the work of Jean Pierre Hardy in Pre-Industrial Technology in the province of Quebec, and Dave Richeson's investigation into telecommunications. The division moreover is also providing an important service in the publication of the *Material Culture Bulletin*, which often contains articles of interest and relevance to the history of science and technology in Canada.

Two sections within the Ontario Ministry of Culture and Recreation have the responsibility for directing and planning the government's historical programming and most historical matters with the government. The Heritage Conservation branch is primarily concerned with particular site development. Included in this are sites such as Ste. Marie-among-the-Hurons, the Naval Establishment at Penetanguishene, Fort William outside Thunder Bay, and the provincial historical plaque programme. In both instances research which may be relevant to the history of science and technology is carried out. The Historical Planning and Research branch is concerned with the broader areas of history and government. In particular this group is charged with supplying historical material where such is required under the Environmental Protection Act, and by the Niagara Escarpment Commission, and other government departments for the review of municipal official plans. Work has been done on the Welland Canal, mining at Cobalt, lime manufacturing on the Niagara Escarpment, and small-scale mining on the South Shield. This and other research is available at the libraries of the Ministries of Natural Resources of Culture and Recreation.

At Upper Canada Village a systematic indexing of the newspapers of eastern Ontario, of the pictorial collections of major public institutions in Ontario, and of agricultural

periodicals is under way. These are proving very useful
in the documentation of technological development in
eastern Ontario. Besides these general projects a study
of the printing trade in eastern Ontario has been completed.

Few of the remaining museums in Central Canada identified on the accompanying sheets are active in research.
The absence of adequate funding means that curators and
others can provide little more than information for each
artifact in their collections. For example, even the new
Ontario Agricultural Museum at Milton has no full-time
research staff at all. The librarian is valiantly attempting to catalogue the mountains of artifacts. In its collection there are significant examples of technological
development in threshing machines, reapers, and tractors
and in its archives are the papers of the International
Harvester Company, among other items, and much of the
Ontario Ministry of Agriculture's historical material.
The Marine Museum of Upper Canada at Toronto has a substantial collection of items related to shipping on the
Great Lakes. But it has no research staff and no prospect
of hiring any.

The Oil Museum of Canada has a small collection of
artifacts and manuscript material. Not only has the museum
no research staff but its Board has actively frustrated a
researcher's attempts to work on their material.

The Pump House Steam Museum in Kingston has an
interesting collection of steam engines and the massive
pumping equipment for the Kingston water works; but for
research, the museum relies on the goodwill of the faculty
and students in the Engineering Department of Queen's
University.

Students with an interest in the history of medicine
in Canada cannot afford to miss the Academy of Medicine
Museum in Toronto. Its collections are unsurpassed in
this country. Unfortunately budgetary restraints and
other considerations make it necessary for users of the
collection to make appointments before they arrive at the
front door. A significant collection of railway cars and
locomotives in Canada has been assembled by the Canadian

Railroad Historical Association at the Canadian Railway
Museum at Delson, Quebec. Other branches of the association have smaller collections across the country. Its
volunteer supporters are very knowledgeable about aspects
of the collection but they are unequal to the enormous task
of providing protection and conservation for this massive
collection. This museum also has a substantial library
and archives.

Bell Canada has the outstanding collection of telephone equipment in Canada. It is very substantial and presumably well-documented by Bell technicians. Bell Canada
is currently hoping to donate the entire collection so as
to form the nucleus of a National Telecommunications Museum; but progress has been slow in turning this offer into
a reality.

It is evident that in this country there are a number
of important collections on a wide range of subject areas
within the scope of science and technology. Even though
these collections often do not cover the entire field
adequately and several areas have been identified as
remaining largely unrepresented, they do provide the nucleus for further development and study. Unfortunately
these museums are for political or financial reasons not
performing adequate research either on their collections
or on the industries which produced them. Consequently
those of you who are active researchers into the history
of science and technology in Canada and wish to use museum
collections must expect to work on your own. I hope this
paper will serve as a useful starting point for this workshop as well as your subsequent encounters with museum
organization.

COMMENTARY
J. Wardrop

Porter's study concentrates on central Canadian institutions and after a brief review of their work, I would
like to respond to his request and expand on some work
being done in British Columbian museums. Parks Canada's
Historic Sites researches science and technology for the

most simple reason of all--its application to a particular
historic site. Their studies which are usually available
to the museum curator are very informative and should be
acquired by all interested museologists. Another successful institution is the National Aeronautic Museum which,
ironically enough, is under the aegis of the National Museum of Science and Technology. Its collection has captured
worldwide respect and, more importantly, it has carried
out extensive research into their collections with the
attendant publications resulting. However, because it does
not follow the example from its subordinate member, the
National Museum of Science and Technology gets a poor review from Porter. The museum's emphasis on promotion of
its collection is a practice we should question: to what
state can industrial artifacts be preserved or conserved
and, finally, "used" for interpretation? I have read a
great deal of material which professes to illuminate the
reader as to the museum conservator's "code of ethics" and
while it is not clear what their stand is on industrial
artifacts, I suspect they would object to the commonly
accepted standards of artifact restoration. To the purist,
every repair to an industrial artifact to make it operate
or appear complete lessens its authenticity as a three-
dimensional document.

In an attempt to round out the Canadian museum picture, I'd like to review progress in the discipline in
British Columbia. Perhaps because this western province
has four such obviously high-profile industries--forestry,
fishing, mining, and agriculture--it was immediately seen
to be mandatory to direct at least fifty percent of our
research, storage, and exhibit resources towards the history of science and technology in this province. Since
the opening of the Modern History Galleries in 1972, the
B.C. Provincial Museum has gone beyond these four basic
industries, and, for example, developed the Provincial
Museum Train which tells the story of the "Age of Steam"
in British Columbia in three railway coaches and two flat-
cars besides the steam locie pulling the exhibits.

Other museums in British Columbia dealing with science

and technology, with the emphasis on the latter (like most
Canadian museums), include B.C. Forest Museum in Duncan,
B.C. Mining Museum in Brittania Beach, B.C. Farming Museum
in Langley, and the Saanichton Historical Artifacts Society
near Victoria. The four have common features: extensive
artifact collections, growing primary documentation archives, but a minimum of staff available to do the artifact
research. In most cases, the above museums find it necessary to "promote" their collections to sustain public
interest.

Our next step is to realize many of our excuses for
lack of progress must be seen as challenges that can be
overcome. I would like to throw out some suggestions -
they're only that because I'm still looking for answers.
Here I must qualify my following statements: they are intended primarily for the general museum which must use
other criteria than those followed by the specialized museum of science and technology. The greatest hindrance
to the development of a strong collection with supporting
evidence is that most artifact research is done after the
collection has been accessioned--instead, research should
be carried out first in order to determine the significant
artifacts; then they should be collected and the necessary
research carried out. This significance can, I believe,
be determined by adding the simple term "culture" to
science and technology. This is not a new idea--*Technology
& Culture*, a journal published quarterly by the University
of Chicago Press since 1959, has often considered this
concept. Such criteria will help ensure that the posses-
sive or antiquarian instincts in so many of us will be
curbed. If the curator sees the scientific and technological artifacts as documents upon which one can base the
history of a region, province, or nation (that may sound
overly ambitious but isn't), the curator will then collect,
preserve, and exhibit or interpret what advances did have
a significant effect on our development. The responsible
curator would not feel obliged to save a prototype of a
steam engine, however interesting and unique, that was not
used. A record of it or an acknowledgment of its existence

would suffice.

The shortage of space is a challenge that leads the curator to consider other means of preserving and presenting the information--surely that is as important if not more so than saving every artifact. For display, models of large industrial processes complemented by some manageable artifact as visible proof of its existence is a justifiable alternative.

It may appear cynical but the conclusion we, at the B.C. Provincial Museum (BCPM), have come to is that each and every offered or considered artifact, especially industrial, must justify the space and manpower it requires. One example of an industrial artifact selected for display in BCPM, while many others in the same industrial process have remained uncollected although recorded, is a fish-gutting machine. The "Iron Chink," with its unfortunate ethnic slur, is a technological innovation which at once reflects the social, political, and economic climate of turn-of-century British Columbia. The west coast salmon canneries, by 1900, were heavily dependent upon highly efficient gangs of Chinese fish butchers. Although their wages were lower than those of the white men, the Chinese were under constant attack by racist British Columbians. It led to successful lobbying of provincial and federal governments to reduce the flow of immigrant Chinese. This fish-gutting machine, developed by Charles Smith of Seattle, Washington was seen as a means to replace the "yellow horde" in the canneries. In 1905, agitation in legislatures forced the government to impose a $500 head-tax upon incoming Chinese. Indeed, it reduced the inflow. Consequently it meant raising the wages of the remaining Chinese which forced more canneries to adopt the new technological innovation--the "Iron Chink."

As for the lack of a rewarding liaison with industry, this situation can be easily rectified. Norman Ball, the Science and Engineering Archivist of the Public Archives of Canada, notes no shortage of potential industrial donors of wide variety but little time and resources to handle what he terms his "hit list." The preconceived list of what scientific or technological sources to contact indicates the necessary research has been carried out so that most resources are directed towards significant trends. Here, another example from BCPM: in the field of industrial technology, our "hit list" includes significant British Columbia producers. This is partly because our mandate is indeed this province's activities, but also because the knowledge of the province's producers, whether they be grist millers or logging equipment manufacturers, greatly enhances the museum's understanding of our experience.

In closing, it becomes clear that the ball is in our court and it is incumbent upon us, the participants in this workshop, to lobby for a journal which would contain the academic spade work so sorely lacking. And it is incumbent upon us to write those scholarly articles necessary to provide a foundation for the study of the history of science and technology for Canadian museologists.

DISCUSSION

Arnold Roos states that much of the preliminary discussion in the workshop was the enumeration of the weaknesses of the National Museum of Science and Technology (NMST), particularly its lack of interpretation and its refusal to do historical studies, to publish, or to build the collection in a systematic way. Some of the same objections were raised over the many small, specialized collections in Canada, but it was noted that museums have to take into account their prospective audiences and act accordingly.

While there was concern voiced about the weakness of museology programmes in times of restraint, a suggestion was made that museums might better employ teams whose members had a variety of expertise areas. The workshop also heard reports on the federal government's assistance programmes to specialized museums, but was told that few museums request the monies available, possibly due to the lack of publicity, partly due to the fact that many museums lack trained personnel.

The consensus was that museums for science and technology will only improve if public pressure is brought to bear upon governments and that this is best done by individuals and the media. The idea was put forth that the media and individuals would continue to function as sleeping watchdogs and that the natural embarrassment caused by the policies and poor funding and performance of the NMST was just the most glaring example of the anti-scientific and anti-technological bias in Canadian "cultural" circles.

A.8: HISTORY OF MEDICINE/L'HISTOIRE DE LA MEDECINE

REPORT ON THE STATUS OF HISTORY OF MEDICINE IN
CANADIAN UNIVERSITIES

Toby Gelfand

The following report is based on the responses to a
survey questionnaire sent to the Deans of Medical Faculties of nine Canadian universities in September 1978.
Four questions were asked:

(1) Does your university offer instruction in the History
of Medicine?

(2) Does your university offer courses in the History of
Science?

(3) How would you rate your library holdings in the area
of the History of Medicine?

(4) Do you think there would be interest in developing
an academic program in this field?

Before looking at the results of the questionnaire,
several qualifications need to be made. First, the survey
makes no claim to be a comprehensive in-depth study. It
invited and obtained brief, sometimes one-word, responses.
Second, the questions and replies are qualitative and, to
an extent, subjective. This should be borne in mind when
considering the various responses. Third, since the survey
was addressed only to the Deans of Medical Faculties, many
potential sources of information have not been tapped.
Eight of the nine Deans did respond (with the exception of
Université de Montréal), but they may have overlooked activities in the history of medicine in other faculties of
their universities. For example, the Department of History at Laval has shown an interest in the history of
medicine. Also, this survey does not take into account
activities in the field in nonmedical universities, e.g.,
York, Carleton, Concordia. Finally, the survey does not
include endowed chairs of History of Medicine.

The results of the survey may be summarized as
follows:

(1) All respondents described their library holdings
in the history of medicine as weak or average.

(2) Despite a bias towards history of medicine in the
survey, the teaching of history of science appeared
to be somewhat wider and presumably more securely
implanted in the curricula than the teaching of
history of medicine. (Five reported history of science
as opposed to three, or possibly four counting Laval,
reporting history of medicine.)

(3) The status of the history of medicine fell
roughly into three configurations based on the number
of positive replies to the four questions posed.
Calgary, Memorial, and Saskatchewan show a strong
configuration (three positive replies). Manitoba and
Laval replied positively to two of the four categories. And Dalhousie, Alberta, and Sherbrooke constituted the weakest configuration, each giving only one
positive reply.

(4) "Interest" was conspicuous on the part of all
respondents, except perhaps Dalhousie. Positive
interest contrasted with average or weak library
resources in three cases (Calgary, Memorial,
Sherbrooke).

It would indeed be hazardous to venture any conclusions based on a survey as sketchy as this one. However,
certain tentative implications can be drawn. Further, I
shall make several recommendations which, though they are
consistent with the above findings, reflect my personal
opinions. These remarks fall into three areas: library
resources, teaching, and research.

First the survey confirms, if it needed confirming,
that interest in the history of medicine exceeds available
library resources. One would, therefore, be meeting a
real demand if library development in this field could be
encouraged. It would seem a simple enough matter to provide the universities with a checklist of essential materials in the field, such as key reference works,

bibliographies, and periodicals. The Hannah Institute is currently engaged in generating and diffusing such materials and perhaps the Canadian Society for the History of Medicine could also play a part in this endeavour. A realistic goal might be to have every institution provided with the two or three major bibliographical tools; e.g., *The Surgeon General's Index*, The NLM *Bibliography of the History of Medicine*, *Catalogue* of the Osler Library, etc., and subscriptions to major journals in the history of medicine.

With regard to the teaching of the history of medicine, some conclusions may be drawn from the successful implantation of the history of science in undergraduate curricula. History of medicine would do well to follow the example of directing its teaching beyond the medical faculty to undergraduate and graduate students in other sectors of the university. The favourable reception of history of medicine at the undergraduate level (drawing upon the large premedical student demand) has been observed by most teachers of the subject and was one of the main conclusions of a recent Macy Foundation study of the teaching of the history of medicine. Specifically, students from departments of biology, philosophy, history, and various social sciences have displayed much interest in the field. Undergraduate interest can best be stimulated by a vigorous program of research in the history of medicine. Active research by professors and at the graduate level would seem to be one of the keys to promoting teaching of the history of medicine at all levels.

A *centre* for research in the history of medicine may be the best way to meet the increased interest and demands in this area. I recommend that thought be given to setting up such a centre which would establish priorities for research on certain themes likely to attract researchers and funds. Two examples of projects which come to mind are (1) history of public health in Canada, including the role of epidemics and the response of medical institutions such as hospitals, and (2) the history of the medical profession in Canada. Regardless of which specific projects are

pursued, I think it important that the research centre
commit itself to certain definite projects and that these
contain the following components: (1) Canadian social
history, (2) history of science and technology, (3) substantial use of archival materials.

The centre could include a permanent staff consisting
of the five Hannah Professors in the history of medicine,
the visiting Hannah Professor, the five posts available
for postdoctoral research connected with the Hannah chairs,
and a number of rotating academic appointments in relevant
fields associated with the history of medicine.

What I am suggesting is coordination of structures
presently existing or potentially available rather than
the creation of anything totally new. The function of the
proposed centre would be (1) to proceed with research on
a specific project with a view toward eventual publication
and (2) training of undergraduate and graduate students
in connection with the project. It would be important to
develop the mechanisms and means for *movement* of professors
and students so that resources already available could be
more efficiently used. The economies of such geographical
movement as compared with establishment of new positions
are obvious. The success of any program designed to
increase research in the history of medicine, of course,
hinges on the broader economic situation. In particular,
one would not want to train students at the graduate level
without regard to career possibilities. I suggest that
the market for skills in the history of medicine beyond
academic work needs to be further explored. Work in archival, museum, and publishing fields might be a favourable
consequence of increased research in the history of medicine. In any case, it would seem, in conclusion, that the
stimulation and coordination of scholarly research in the
field of Canadian history of medicine is a necessary first
step toward the expansion of this field.

THE MEDICAL ARCHIVES INVENTORY PROJECT
Margaret Dunn
The Hannah Institute for the History of Medicine,

incorporated as a nonprofit institution, actively seeks
to promote interest in the history of medicine through
grants-in-aid for study and research work, publications
assistance, awards, and Hannah Lectures by distinguished
historians.

The Jason A. Hannah Chairs for the History of Medicine, now established at each of the five medical faculties
in the province of Ontario, provide a focal point for encouragement of academic pursuit in the field of the history of medicine for students of the profession as well
as for others enrolled in social sciences or humanities
disciplines. Any individual beginning work in a new field
of interest has reason to want to consult good bibliographies of secondary source materials. But in Canada,
in the past, little work has been carried out in preparation of comprehensive annotated bibliographies of
Canadian secondary source materials, and few have been
aware of the existence and conditions of primary source
materials.

The first need should be admirably met in the near
future with Dr. Charles Roland's bibliography and his col-
laborative effort with Dr. Paul Potter on a bibliography
of Canadian medical journals. A further aid will be Gail
Moore's bibliography of medical literature in the Robarts
Research Library of the University of Toronto. Once the
secondary source materials are more generally known,
interest in advanced study can be reasonably assured if
archival materials are known to be in good condition and
available to researchers. The Symons report *To Know
Ourselves* called attention to the importance of archival
collections for original research in a variety of areas
of Canadian studies. Unfortunately, in far too many instances over the past several decades, archives suffered
as a low priority in administrative hierarchies and where
they did exist, it was frequently thanks to the perse-
verance of a dedicated few.

In 1978 then, the Hannah Institute Board of Directors
agreed that an information base of primary source materials
relating to medicine should be prepared for the use of

scholars and students interested in studying some aspect
of medical history, but desirous of guidance as to sources
of materials. A researcher was hired for a two-year period
with the task of locating, identifying, and preparing a
directory of the medical archives collections within the
province.

This undertaking had a number of distinguishing features: it was a natural area of interest for the Hannah
Institute and could be promoted within the established
lines of communication; it would ideally bring to light
a vast amount of hitherto unknown information regarding
largely unexplored areas of medical activity within this
province; researchers consulting the directory would be
provided with an estimation of the extent and scope of the
collections and could so gauge the potential applicability
and relevance of certain documentation for their research;
the project would provide a good focal point in approaches
to medical practitioners and institutions and alert them
to the importance of archives, whether as a local interest
pastime or as a professional undertaking; significant
savings of heritage materials presently in the hands of
individuals or institutions might be realized if suggestions on donations to appropriate archival repositories
were taken up.

Cooperation and collaboration are the bywords in en-
suring the success of this project. A singular benefit
for the Hannah researcher is the collective knowledge and
counsel of the professional staff of the Public Archives
of Canada and the Archives of Ontario. For several years
now staff at the Public Archives have been involved in a
regular updating of the bible of archives: the *Union List
of Manuscripts/Catalogue Collectif des Manuscrits*. This
union list serves as a first-step identification of some
of the medical manuscript material held by reporting institutions across Canada. Also noteworthy is the Toronto
Area Archives' Group work on a projected fifteen-volume
series entitled *Ontario's Heritage: A Guide to Archival
Sources*. The geographic approach will mean that in time
there will be virtually complete access to all types of

records for a host of institutions and agencies in this
province.

The parameters of the Hannah medical archives project
are presently extremely broad. The research may be
likened to a fishing trip--there is no preknowledge of the
catch. Hence the researcher must anticipate including
materials of practising professionals, health sciences
associations, university medical faculties, government departments, hospitals, clinics, laboratories, and social
service agencies such as Houses of Industry whose activities were frequently the precursors to medical establish-
ments as we know them today.

Social history is certainly one field where medical
archival materials can be used to advantage, but other
disciplines can equally well interplay in a study: public
administration, political science, geography, psychology,
sociology, anthropology, architecture, etc. Records of
government departments alone yield an incredibly diverse
array of material ranging from occupational health to
patent medicines. In staking out this unknown territory,
the researcher spent varying periods of time between
August and November of 1978 working in Toronto, Ottawa,
and Kingston. Holdings of medical archives collections
have been noted for the following institutions to date:
Public Archives of Canada, National Research Council, Medical Research Council, Canadian Public Health Association,
Canadian Medical Association, Canadian Nursing Association,
Archives of Ontario, Academy of Medicine at Toronto,
St. Michael's Hospital at Toronto, and Queen's University
Archives.

Notations on these collections and suggestions and
advice from other professionals will be used as a basis for
discussions regarding alterations or refinements in defin-
itions, methodology, itinerary, etc. Equally important is
the necessity of a professional public relations approach
to persuade all those individuals and institutions identified as medical that discovering the significance of the
materials in their possession can prove as important and
gratifying to them as to the outside research community.

THE HISTORY OF MEDICINE MUSEUM

John W. Scott

On this continent museums illustrating the history
of medicine fall into two main groups: those whose prime
function is to illustrate the social history of a region
and those which try to demonstrate the progress of the
art and science of medicine through the ages and its influence on the social order. The Ontario Science Centre
in Toronto represents a special variant of the second
group where the progress of medicine is portrayed in the
context of the overall development of science and technology.

Medicine is deeply involved in the social structure
of society and the medicine of a period reflects the be-
liefs and philosophy of that period. So it is proper that
many small local history museums have displays of the instruments used by the first practitioner in their area, or
the pill, tile, and stock bottles from the first pharmacy.
Perhaps there is a photograph of the opening of the hospital or the installation of the first x-ray unit. These
displays give a glimpse of the state of the art at a specific time, but usually make no attempt to illustrate the
progress of medicine over the past two centuries.

The displays in the restored forts, e.g., Old Fort
York in Toronto, often display the surgeon's tools. At
the time of the Rebellion in 1837 amputation was the
standard treatment for serious wounds, and there was no
elective surgery because of the inevitable infection. The
amputation kit contains long knives, a saw, and a tourni-
quet. Usually there is also a medicine chest with its
bottles, balance, and mortar.

In some towns, as in Peterborough, Ontario, a prominent early doctor's house has been restored in conformity
with the social context at a specific time. Similarly,
in Upper Canada Village and Black Creek Pioneer Village
the Doctor's Residence is represented. At Black Creek
the period selected was 1867 and his consulting room is
furnished accordingly with a desk, a cabinet of drugs for
dispensing, surgical instruments, an examining table which

doubled as an operating table, and a few books. In the
social context the history of medicine is shown as it was
at a specific time in a specific social environment, i.e.,
a "cross section."

At the History of Medicine Museum of the Academy of
Medicine, Toronto we are attempting to give a "longitudinal
view" of medical progress. This is only partially
achieved, because our collection is limited, although I
believe it is the largest in Canada. The size and scope
of such a museum is almost limitless for anything dealing
with the health, disease, or injury from prehistoric
times to the present properly belongs. As a result most
history of medicine museums limit themselves by setting
rigid guidelines limiting their collection.

The Armed Forces Medical Services Museum at Camp
Borden has limited itself to artifacts pertaining to the
provision of medical care for Canadian Forces. The museum
at the Hôtel-Dieu in Quebec City records the founding and
growth of that hospital. The Heart Surgery Museum at the
Toronto General Hospital illustrates the progress of
open-heart surgery during the last thirty years.

At the Academy of Medicine we have not limited our
field, but special emphasis has been given to the history
of medicine in the Greater Toronto Region, and to paleopathology. Consequently we have displays relating to the
Indian Shaman; military medicine from 1776 to the present;
pioneer medicine in Ontario; the contribution of Sir
William Osler; the advances made by the Toronto medical
schools including insulin, Heparin, Cyclopropane, surgical
instruments, the Electrocardiogram, the Anti-G suit, etc.
There is a growing collection of Paleopathology and a
scattering of Greek, Roman, Chinese, and post-Renaissance
European artifacts. The Drake Collection dealing with all
aspects of the care and feeding of infants and children
throughout the ages, gives an indication of medical progress
over the past twenty-five centuries.

In effect the History of Medicine Museum provides
source material for the historian of medicine and for the
social historian.

THE POSSIBLE CONSEQUENCES OF A BIBLIOGRAPHY
OF CANADIAN MEDICAL PERIODICALS

C. G. Roland

Over the past two years we have completed the manuscript of a bibliography of Canadian medical journals (C. Roland and P. Potter, *An Annotated Bibliography of Canadian Medical Journals, 1826-1975*, in preparation). Two hundred and nine titles are identified as having begun publication during the one hundred fifty years covered by the study. The imminent availability of this compilation suggests a number of potential consequences, by no means all of which are desirable.

Intellectual and Physical Effects

These possible effects can be categorized as intellectual and physical. The intellectual consequences, described in broad terms, are two: (1) there are the various unspecifiable scholarly uses that one hopes and expects will accrue from any bibliographic project; also, (2) libraries and collectors should be aided in assessing holdings and needs. These presumed effects seem entirely beneficial even if attained only to a modest degree.

The physical effects may be less benign; they include consequences related to availability and to increased use.

(1) *Availability*: the nineteenth-century Canadian medical periodicals are by no means widely available, particularly those published before 1900. Of the two hundred nine titles identified, the National Library of Medicine in Bethesda, Maryland holds either the sole extant copy or the only (or best) complete run of eight titles; a further thirteen titles are noted in the bibliography as "not located." This last group contains journals described by presumably reliable sources as having existed, although we have been unable to find copies in any library. Thus ten percent of the titles on our list are not available anywhere or are unavailable in Canada.

At the very least, this consequence is an embarrassment and a mild hindrance to scholarship. Offsetting this effect is the fact that the scarce or vanished journals are all short-run publications that are, therefore, probably of limited importance.

(2) *Increased Use*: If the bibliography serves its function properly, it will produce among scholars a greater awareness of the scope of Canadian medical journals, especially the uncommon and short-run journals. This effect, coupled with the growth of activity in Canadian medical history generally, will mean that the journals will be used more often by more readers. But most Canadian libraries have scanty holdings of these journals. The most complete collections are those of the William Boyd Library of the Academy of Medicine of Toronto and the medical libraries of McGill University and the University of Western Ontario.

The nineteenth- and early twentieth-century journals typically have fared badly with the passage of time, many of them having been printed on paper that has become extremely brittle. Thus with increased usage the damage to journals will become more and more severe, in a cycle that could end in restrictions on usage and especially on copying privileges.

Additionally, there is the reasonable expectation that when any of these journals does come on the market, the increased demand will be reflected in correspondingly higher sale prices.

Solutions ?

Thus I foresee desirable intellectual consequences but potentially serious physical damage as products of our bibliography of Canadian medical periodicals. What is the solution? Certainly not the negative one of suppressing our bibliographic manuscript.

I would suggest that an ideal response to these expressions of concern about potential physical damage would be a comprehensive program to prepare and distribute microform copies of all two hundred nine journals, or at least those of the pre-World War I component. These journals are extremely important primary sources for Canadian medical and general historians. Once microfiche are made, the use of the fragile originals can be restricted, yet the material will be available for study and potentially would be much more widely available than is the case now because libraries across the country could own complete sets of these valuable historical tools.

Thus a second-generation consequence of our bibliography could be a proposal to persuade appropriate governmental agencies or private institutions to undertake such a microfilming project.

DISCUSSION

Since the history of medicine session consisted of four short papers, there was no commentary. Wendy Mitchinson reported that there are two central problems in the history of Canadian medicine, that of teaching and research and that of resources. There is little indication of the extent to which history of medicine is being researched. Evidence presented in the session pertained to teaching in medical schools, but there is no information on the interest of nonmedical students in the subject. The participants heard about research into professionalization and institutionalization of Canadian medicine but there was a lack of discussion on folk medicine and other aspects of the field. The group saw the lack of funding as a major weakness in the expansion of the teaching of the history of medicine.

Resources for the history of medicine are, on the other hand, quite good in Canada. Archives, library collections, museums, journal runs, etc., are in place but all require extensive funding for maximum use by the researcher and the public. While no resolutions came out of the session, the group felt that it is time for extensive stocktaking of resources, both human and material.

The history of Canadian medicine has advantages over that of either science or technology. It has a relatively clear-cut subject matter, professionals in endowed chairs in Ontario, a national society with a strong Canadian bias, and, above all, an active funding agency in the Hannah Institute. The talk of the lack of funds is hardly to the point for this field; rather, it is the lack of manpower. Since funds are available, archival resources good, and

the inventory of resources well under way, the issue of
coordination, raised by Gelfand, should be addressed
seriously by those in the field.

The history of medicine has always fascinated physi-
cians, perhaps even more so than history has excited
scientists and engineers. The professional historians,
linked with the Canadian Society for History of Medicine
and the Hannah Institute, should look for ways to utilize
this potentially valuable human resource pool. If doctors
and nurses wish to write history, then the professionals
could offer them assistance by means of short courses,
consulting services, and a central clearinghouse for
materials. The suggestion raised in the teaching materials
workshop on a central registry of autobiographical works
would apply equally well to those in the health sciences.
Such a national registry (or even depository) and the encouragement of doctors to write could be coordinated
easily, especially if the medical societies were enlisted
in the project.

APPENDIX B

APPENDICE B

MUSEUMS OF CANADIAN SCIENCE

AND TECHNOLOGY

MUSEES DES SCIENCES ET DE LA

TECHNOLOGIE CANADIENNES

Christopher J. B. L. Porter

MUSEUMS OF CANADIAN SCIENCE AND TECHNOLOGY/
MUSEES DES SCIENCES ET DE LA TECHNOLOGIE
CANADIENNES

This list is in two parts, the first noting the major scientific and technical collections in Canada, the second being an outline of the areas of research presently being undertaken by Canadian museums. For further information, the reader should consult: Barry Lord, *Specialized Museums in Canada:A Report to the Museum Assistance Programmes of the National Museums of Canada* (Ottawa, 1977), and also, Canadian Museums Association, *Directory of Canadian Museums* (Ottawa, 1976).

MUSEUMS/MUSEES

INDUSTRIAL

Agriculture
1. National Museum of Science and Technology, Ottawa
2. Ontario Agricultural Museum, Milton, Ont.
3. Western Development Museum, Moose Jaw, North Battleford, Saskatoon, Yorkton, Sask.
4. British Columbia Farm Machinery Museum, Langley, B.C.
5. Manitoba Agricultural Museum, Austin, Man.
6. Ross Farm Museum of Agriculture, New Ross, N.S.
7. Reynolds Museum, Wetaskiwin, Alta.

Energy
1. Pump House Steam Museum, Kingston, Ont.
2. Hamilton Water Works, Hamilton, Ont.

Forest Industries
1. British Columbia Forest Museum, Duncan, B.C.
2. Logging Museum, Algonquin Park, Ont.
3. Sawmill, Upper Canada Village, Morrisburg, Ont.
4. Sawmill, Moira Conservation Authority, Madoc, Ont.

Metal Manufacturing
1. Forges St.-Maurice, Trois-Rivières, Qué.

Mining
1. Cape Breton Miners Museum, Glace Bay, N.S.
2. Stellarton Miners Museum, Stellarton, N.S.
3. Springhill Miners Museum, Springhill, N.S.
4. British Columbia Museum of Mining, Britannia Beach, B.C.
5. Cobalt Northern Ontario Mining Museum, Cobalt, Ont.

Textiles
1. Upper Canada Village
2. Dean Wile Carding Mill, Bridgewater, N.S.
3. Old Woolen Mill Museum, Barrington, N.S.
4. Ontario Science Centre, Don Mills, Ont.

MEDICAL

1. Academy of Medicine Museum, Toronto, Ont.
2. McGill University Medical Museum, Montréal, Qué.

TRANSPORTATION & COMMUNICATIONS

Automobiles
1. National Museum of Science and Technology
2. Classic Car Museum, Victoria, B.C.
3. Manitoba Automobile Museum, Elkhorn, Man.
4. Canadian Automotive Museum, Oshawa, Ont.
5. Reynolds Museum
6. Western Development Museum

Aviation
1. National Aeronautical Collection

Communications-Telegraph and Telephone
1. Bell Canada Museum, Montréal, Qué.
2. Alberta Government Telephone, Edmonton, Alta.
3. New Brunswick Telephone, N.B.
4. Maritime Telephones and Telecommunications
5. National Museum of Science and Technology

Communications-Printing
1. National Museum of Science and Technology
2. Upper Canada Village
3. Black Creek Pioneer Village, Downsview, Ont.

Maritime/Fisheries/Great Lakes
1. Lunenburg Fisheries Museum, Lunenburg, N.S.
2. Maritime Museum of Canada, Halifax, N.S.
3. Maritime Museum of British Columbia, Victoria, B.C.
4. Vancouver Maritime Museum, Vancouver, B.C.
5. Marine Museum of Upper Canada, Toronto, Ont.
6. Maritime Museum of Manitoba, Selkirk, Man.
7. Kanawa International Museum, Minden, Ont.

Railways
1. Canadian Railway Museum, Delson, Qué.
2. National Museum of Science and Technology

CURRENT RESEARCH/RECHERCHE ACTUELLE

Parks Canada
1. British military engineering adaptations to the
   Canadian environment (especially Halifax)
2. Landscape history bibliography and reference file
3. Technological and innovative changes in ordnance
   1760-1900--supply, manufacture, design
4. Servant classes in the Hudson's Bay Company
5. Loyalists--site-related studies
6. Francophone minorities--Acadians
7. Domestic servants in Québec and Maritimes
8. British Army trades
9. Bibliography of household manuals

10. Domestic Life and domestic accoutrements in
    Québec
11. Provincial Marine, especially Niagara frontier
12. 19th century Canadian wallpapers
13. Alexander G. Bell artifacts and manuscripts, Bell
    Museum, Baddeck, N.S.
14. Gold mining technology in the Yukon
15. S.S. Klondike and Yukon steamboats
16. Grist and flour mills in Ontario, 1780-1880
17. Shipbuilding on the Rideau Canal
18. Commercial navigation, road and rail transport
    in Rideau corridor
19. Canal building technology, Rideau
20. Pictorial inventory of Rideau
21. Agriculture and industry in Rideau corridor
22. Encyclopedia of transportation (planned)
23. Documentation of Forges St.-Maurice
24. Documentation of buildings, life, and clothing
    of Louisbourg
25. Canning industry in Revelstoke
26. Publications: *Bulletin, Manuscript Report* series
27. Canadian Inventory of Historical Buildings. Use-
    ful tool for architecture and style.

Upper Canada Village
1. Industrial and agricultural history of Eastern
   Ontario with newspapers
2. Pictorial inventory of Ontario (1760-1875) compiled
   from Ontario and Montréal institutional sources.
3. Printing in Eastern Ontario, 1830-1867

Royal Ontario Museum/University of Toronto Museology Programme
1. 19th-century sewing machines
2. 19th-century pianos
3. Catalogue of historic scientific instruments of
   the University of Toronto

National Museum of Man-History Division
1. Pre-industrial technology in Québec
2. Collection inventory of major Canadian museums
   (computerized inventory)
3. Electric Telegraph, 1846-1902
4. Manufacture of porcelain and china
5. Ontario furniture and cabinet makers
6. 19th-century costume
7. Food processing in Ontario
8. Publications: *Material History Bulletin, Mercury
   Series, Collection histoire populaire du Québec*

British Columbia Provincial Museum
1. Pacific canning industry
2. Forest industry
3. Industrial support industries--foundries, coal
   mining, and packaging

APPENDIX C

APPENDICE C

SELECT BIBLIOGRAPHY OF THE HISTORY
OF CANADIAN SCIENCE, MEDICINE, AND TECHNOLOGY

BIBLIOGRAPHIE SELECTIVE D'HISTOIRE
DES SCIENCES, DE LA MEDECINE ET DE LA TECHNOLOGIE
CANADIENNES

Richard A. Jarrell and *Arnold E. Roos*
with the assistance of
W. G. Richardson and *P. J. Bowler*

SELECT BIBLIOGRAPHY/BIBLIOGRAPHIE SELECTIVE

The bibliography comprises some 250 titles which represent the more important items of the several thousand collected by the principal compilers. For the most part, journal literature is ignored for the published monographs. Naturally, the quality of these items varies considerably. The classification of subject matter is broad and the listing is alphabetical. Those wishing additional bibliography should consult the *Bibliography for Courses in the History of Canadian Science, Medicine, and Technology* by Jarrell and Roos (published by HSTC Publications).

A. RESEARCH TOOLS/OUTILS DE RECHERCHE

Beaulieu, A., J.-C. Bonenfant, J. Hamelin. *Repertoire des Publications gouvernmentales du Québec de 1867 à 1964*. Québec, 1968.

Beaulieu, A., J. Hamelin, B. Bernier. *Guide d'histoire du Canada*. Québec, 1969.

Bishop, O. B. *Publications of the Government of the Province of Canada, 1841-1867*. Ottawa, 1963.

Harris, R. S., and A. Tremblay. *Bibliography of Higher Education in Canada*. Toronto, 1960. *Supplement*, 1965.

Henderson, G. F. *Federal Royal Commissions in Canada, 1867-1966:A Checklist*. Toronto, 1967.

Jarrell, R. A., and A. E. Roos. *A Bibliography for Courses in the History of Canadian Science, Medicine, and Technology*. Thornhill, Ont., 1979.

McTaggart, H. I. *Publications of the Government of Ontario, 1901-1955*. Toronto, 1964.

Morgan, H. J. *Bibliotheca Canadensis*. Ottawa, 1867.

Richardson, W. G. *A Survey of Canadian Mining History*. Montreal, 1974.

Thibault, C. *Bibliographia Canadiana*. Toronto, 1973.

B. GENERAL WORKS/OUVRAGES GENERAUX

Beals, C. S., and D. Shenstone. *Science, History and Hudson's Bay*. Ottawa, 1968.

Brown, J. J. *Ideas in Exile:A History of Canadian Invention*. Toronto, 1967.

Eggleston, W. *Scientists at War*. Toronto, 1950.

Levere, T. H., and R. A. Jarrell, eds. *A Curious Field-book:Science and Society in Canadian History*. Toronto, 1974.

Lortie, L. "La trame scientifique dans l'histoire du Canada," in G. F. G. Stanley, ed. *Pioneers of Canadian Science*. Toronto, 1966.

Ouellet, C. *la vie des sciences au Canada français*. Québec, 1964.

Roy, A. *les lettres, les sciences et les arts au Canada sous le régime français*. Paris, 1930.

Royal Society of Canada. *Fifty Years' Retrospect*. Toronto, 1939.

Sinclair, B., N. Ball, and J. Petersen. *let us be Honest and Modest:Technology and Society in Canadian History*. Toronto, 1974.

Stanley, G. F. G. , ed. *Pioneers of Canadian Science*. Toronto, 1966.

Tory, H. M. ed. *A History of Science in Canada*. Toronto, 1939.

C. SCIENCES

Audet, L.-P. "Hydrographes du Roi et cours d'hydrographie au College de Québec," *Cahiers des Dix* 35 (1970), 13-37.

Douglas, A. V. "The St. Helena Observatory and Canadian Astronomy," *Queen's Quarterly 78* (1971), 592-601.

Eggleston, W. *Canada's Nuclear Story*. Toronto, 1965.

Gray, F. W. "Pioneer Geologists of Nova Scotia," *Dalhousie Review 26* (1947), 467-70.

Jarrell, R. A. "The Birth of Canadian Astrophysics: J. S. Plaskett at the Dominion Observatory," *Journal of the Royal Astronomical Society of Canada 71* (1977), 221-33.

Jarrell, R. A. "Origins of Canadian Government Astronomy," *Journal of the Royal Astronomical Society of Canada 69* (1975), 77-85.

Kennedy, J. E. "Our Heritage in Canadian Astronomy," *Journal of the Royal Astronomical Society of Canada 66* (1972), 83-98.

Le Bourdais, D. M. *Canada and the Atomic Revolution*. Toronto, 1959.

Lortie, L. *Le Traité de chimie de J.-B. Meilleur.* Montréal, 1937.

Middleton, W. E. K. *Physics at the National Research Council of Canada, 1929-1952.* Waterloo, 1979.

Neale, E. W. R. *The Earth Sciences in Canada: A Centennial Appraisal and Forecast.* Toronto, 1968.

O'Brien, C. F. "Eozoön Canadense: 'Dawn Animal of Canada,'" *Isis 61* (1970), 206-23.

Roome, P. "The Darwin Debate in Canada: 1860-1880," in L. A. Knafla, et al., eds. *Science, Technology and Culture in Historical Perspective.* Calgary, 1976, 183-205.

Rousseau, J. "La botanique Canadienne à l'époque de Jacques Cartier," *Annales* ACFAS 3 (1937), 151ff.

Rousseau, J. "Le voyage d'André Michaux au lac Mistassini en 1792," *Mémoires Jardin botanique de Montréal 3.* Montréal, 1948.

Royal Astronomical Society of Canada. *Astronomy in Canada:Yesterday, Today and Tommorrow.* Toronto, 1967.

Thomson, D. W. *Men and Meridians.* 3 vols. Ottawa, 1966-69.

Thomson, M. M. *The Beginning of the Long Dash: A History of Timekeeping in Canada.* Toronto, 1978.

Warrington, C. J.,and B. T. Newbold. *Chemical Canada: Past and Present.* Ottawa, 1970.

Warrington, C. J., and R. V. V. Nicholls. *A History of Chemistry in Canada.* Toronto, 1949.

Young, E. G. *The Development of Biochemistry in Canada.* Toronto, 1976.

D. SPECIFIC TECHNOLOGIES/LES TECHNIQUES

Pioneer, Building, and Surveying Technology/Techniques des pionniers, la construction et l'arpentage

Bond, C. J. J. *Surveyors of Canada, 1867-1967.* Ottawa, 1967.

Broadfoot, B. *The Pioneer Years:1895-1914. Memories of Settlers who Opened the West.* Toronto, 1976.

Crichton, V. *Pioneering in Northern Ontario.* Belleville, 1976.

Douville, R., and J.-P. Casanova. *Daily Life in Early Canada*. London, 1968.

Glover, R. (ed.). *David Thompson's Narrative 1784-1812*. Toronto, 1963.

Goulding, W. S. (ed.). *Historic Architecture of Canada*. Ottawa, 1966.

Gowans, A. *Building Canada:An Architectural History of Canadian Life*. Toronto, 1966.

Guillet, E. C. *Pioneer Arts and Crafts*. Toronto, 1968.

Guillet, E. C. *Early Life in Upper Canada*. Toronto, 1933.

Lessard, M., et G. Vilandré. *La maison traditionelle au Québec: Construction, inventaire, restauration*. Montréal, 1974.

Lower, A. R. M. *Settlement and Forest Frontier in Eastern Canada*. Toronto, 1936.

Moogk, P. *Building a House in New France*. Toronto, 1977.

Parson, J. E. *West on the 49th Parallel: Red River to the Rockies, 1872-1876*. New York, 1963.

Ritchie, J., et al. *Canada Builds, 1867-1967*. Toronto, 1967.

Russell, L. S. *Everyday Life in Colonial Canada*. London, 1973.

Seguin, R.-L. *La civilisation traditionelle de l''Habitant' au 17e et 18e siècles: Fonds materiels*. Montréal, 1967.

Thomson, D. W. *Skyview Canada: A Story of Aerial Photography in Canada*. Ottawa, 1975.

Transportation and/et Communication

Aitken, H. G. J. *The Welland Canal Company: A Study in Canadian Enterprise*. Cambridge, Mass., 1954.

Armour, C. A., and T. Lackey. *Sailing Ships of the Maritimes*. Toronto, 1975.

Ashley, C. A. *The First Twenty-five Years: A Study of Trans-Canada Airlines*. Toronto, 1963.

Berton, P. *The Last Spike: The Great Railway 1881-1885*. Toronto, 1971.

Berton, P. *The National Dream: The Great Railway 1871-1881*. Toronto, 1970.

Blake, H. W. *The Era of Streetcars and Interurbans in Winnipeg 1881 to 1955*. Winnipeg, 1974.

Bruce, R. V. *Alexander Graham Bell and the Conquest of Solitude*. London, 1973.

Bullock, E. J. *Ships and the Seaway*. Toronto, 1959.

Clayton, H. *Atlantic Bridgehead: The Story of Trans-atlantic Communications*. Toronto, 1969.

Collins, R. *A Great Way to Go: The Automobile in Canada*. Toronto, 1973.

Currie, A. W. *The Grand Trunk Railway of Canada*. Toronto, 1957.

Denison, M. *C.C.M.: The Story of the First Fifty Years*. Toronto, 1946.

Dorion, P. C. *Canadian National Railway*. Saanichton, 1975.

Downs, A. *Paddlewheels on the Frontier: The Story of British Columbia and Yukon Sternwheel Steamers*. Surrey, 1972.

Due, J. F. *The Interurban Electric Railway Industry in Canada*. Toronto, 1966.

Ellis, F. H. *Canada's Flying Heritage*. Toronto, 1961.

Fields, H. M. *The Story of the Atlantic Cable*. New York, 1892.

Glazebrook, G. P. de T. *A History of Transportation in Canada*. 2 volumes. Toronto, 1964.

Green, H. G. *The Silver Dart: The Authentic Story of the Hon. J. A. D. McCurdy, Canada's First Pilot*. Fredericton, 1969.

Greenhill, R. *Early Photography in Canada*. Toronto, 1965.

Guillet, E. C. *The Great Migration: The Atlantic Crossing by Sailing Ship since 1770*. Toronto, 1937.

Guillet, E. C. *The Story of Canadian Roads*. Toronto, 1966.

Guillet, E. C. *The Valley of the Trent*. Toronto, 1957.

Innis, H. A. A *History of the Canadian Pacific Railway*. Toronto, 1971.

Keefer, T. C. *The Canals of Canada*. Montreal, 1894.

Keefer, T. C. *The Old Welland Canal and the Man who Made It*. Ottawa, 1911.

Legget, R. F. *The Canals of Canada*. Vancouver, 1975.

Legget, R. F. *Ottawa Waterway: Gateway to a Continent*. Toronto, 1975.

Manchester, L. *Canada's Aviation Industry*. Toronto, 1968.

Manny, L. *Ships of the Miramichi: A History of Shipbuilding on the Miramichi River, N.B., Canada, 1773-1919*. Saint John, 1960.

Miller, M. G. *Straight Lines in Curved Space: Colonization Roads in Eastern Ontario*. Toronto, 1978.

Molson, K. M. *Pioneering in Canadian Air Transport*. Winnipeg, 1974.

Moogk, E. *Roll Back the Years: History of Canadian Recorded Sound and its Legacy, Genesis to 1930*. Ottawa, 1975.

Morton, H. *The Wind Commands: Sailors and Sailing Ships in the Pacific*. Vancouver, 1975.

Parker, J. P. *Sails of the Maritimes: The Story of the Three- and Four-masted Cargo Schooners of Atlantic Canada, 1859-1929*. Halifax, 1960.

Parkin, J. H. *Bell and Baldwin: Their Development of Aerodromes and Hydrodromes at Baddeck, Nova Scotia*. Toronto, 1964.

Patten, W. *Pioneering the Telephone in Canada*. Montreal, 1926.

Pursley, L. *Street Railways of Toronto, 1861-1921*. Toronto, 1958.

Regehr, T. D. *The Canadian Northern Railway: Pioneer Road of the Northern Prairies, 1895-1918*. Toronto, 1975.

Raby, O. *Radio's First Voice: The Story of Reginald Fessenden*. Toronto, 1970.

Snider, C. H. J. *The Silent St. Lawrence*. Toronto, 1949.

Stevens, G. R. *Canadian National Railways. Sixty Years of Trial and Error (1836-1896).* Toronto, 1960.

Stevens, G. R. *History of the Canadian National Railways.* New York, 1973.

Industries and Power Technology/ Les Industries et l'Energie

Creighton, D. G. *The Empire of the St. Lawrence 1760-1850.* Toronto, 1956.

Dales, J. H. *Hydroelectricity and Industrial Development: Quebec 1898-1940.* Cambridge, 1957.

Denison, M. *The People's Power: The History of Ontario Hydro.* Toronto, 1960.

Donald, W. J. *The Canadian Iron and Steel Industry.* Boston, 1915.

Fauteux, J.-N. *Essai sur l'industrie au Canada sous le régime français.* Québec, 1927.

Howay, F. W., W. N. Sage, and H. F. Angus. *British Columbia and the U.S.: The North Pacific Slope from Fur Trade to Aviation.* New Haven, 1942.

Innis, H. A. ed. *The Dairy Industry in Canada.* Toronto, 1937.

Naylor, T. *The History of Canadian Business, 1867-1914.* 2 volumes. Toronto, 1975.

Nelles, H. V. *The Politics of Development: Forests, Mines, and Hydro Electric Power in Ontario, 1849-1941.* Toronto, 1974.

Palardy, J. *The Early Furniture of French Canada.* Toronto, 1963.

Plewman, W. R. *Adam Beck and the Ontario Hydro.* Toronto, 1947.

Priamo, C. *Mills of Canada.* Toronto, 1976.

Reid, D. J. *The Development of the Fraser River Salmon Canning Industry, 1885-1913.* Vancouver, 1973.

Rouillard, J. *Les Travailleurs du cotton au Québec 1900-1915.* Montréal, 1974.

Shackleton, P. *The Furniture of Old Ontario.* Toronto, 1973.

Tucker, G. V. *The Canadian Commercial Revolution 1845-1851.* New Haven, 1936.

Tulchinsky, G. J. J. *The River Barons: Montreal Businessmen and the Growth of Industry and Transportation, 1837-1853.* Toronto, 1977.

Vaillancourt, E. *History of the Brewing Industry in the Province of Quebec.* Montreal, 1940.

Staples and Resources/ les Ressources et les Produits Principals

Berton, P. *Klondike: The last Great Gold Rush, 1896-1932.* Toronto, 1972.

Cox, T. R. *Mills and Markets: A History of the Pacific Coast Lumber Industry to 1900.* Seattle, 1974.

Fish, R. *Historical Highlights of Canadian Mining.* Toronto, 1972.

Gould, E. *Oil: The History of Canada's Oil and Gas Industry.* Saanichton, 1976.

Hazlitt, W. C. *The Great Gold Fields of Cariboo.* Vancouver, 1974.

Innis, H. A. *The Cod Fisheries: The History of an International Economy.* Toronto, 1940.

Innis, H. A. *The Fur Trade in Canada.* Toronto, 1970.

Innis, H. A. *Settlement and the Mining Frontier.* Toronto, 1937.

Jones, R. L. *History of Agriculture in Ontario, 1613-1880.* Toronto, 1946.

Kilbourn, W. *The Elements Combined: A History of the Steel Company of Canada.* Toronto, 1960.

Le Bourdais, D. M. *Metals and Men: The Story of Canadian Mining.* Toronto, 1957.

Le Bourdais, D. M. *Sudbury Basin. The Story of Nickel.* Toronto, 1954.

Letourneau, F. *L'Histoire de l'agriculture au Canada français.* Montréal, 1950.

Lower, A. R. M. *Great Britain's Woodyard: British America and the Timber Trade 1763-1867.* Montreal, 1973.

Moore, E. S. *American Influence in Canadian Mining.* Toronto, 1941.

Murray, S. N. *The Valley Comes of Age: A History of Agriculture in the Valley of the Red River of the North, 1812-1920.* Fargo, N.D., 1967.

Pain, S. A. *The Way North. Men, Mines and Minerals*. Toronto, 1964.

Reaman, G. E. *A History of Agriculture in Ontario*. 2 volumes. Toronto, 1970.

Rich, E. E. *The Fur Trade and the Northwest to 1857*. Toronto, 1967.

Rich. E. E. *Montreal and the Fur Trade*. Montreal, 1965.

Rickard, T. A. *The Romance of Mining*. Toronto, 1944.

Séguin, R.-L. *L'Equipement de la ferme canadienne aux XVIIe et XVIIIe siècles*. Montréal, 1954.

Shaw, M. M. *Geologists and Prospectors*. Toronto, 1958.

Taylor, G. W. *Timber: History of the Forest Industry in British Columbia*. Vancouver, 1975.

Tessier, A. *Les Forges St. Maurice*. Trois-Rivières, 1952.

Trigger, B. G. *The Huron: Farmers of the North*. New York, 1969.

E. SCIENTIFIC AND TECHNICAL EDUCATION/EDUCATION SCIENTIF-
IQUE ET TECHNIQUE

Audet, L.-P. "La fondation de l'Ecole Polytechnique," *Cahiers des Dix 30* (1965), 149-91.

Audet, L.-P. *Histoire de l'enseignement au Québec, 1608-1971*. Montréal, 1971.

Clark, A. L. *The First Fifty Years: a History of the Science Faculty at Queen's, 1893-1943*. Kingston, 1944.

Craigie, E. H. *A History of the Department of Zoology at the University of Toronto up to 1962*. Toronto, 1966.

Firestone, O. J. *Industry and Education: a Century of Canadian Development*. Ottawa, 1969.

Gosselin, A. *L'Instruction au Canada sous le régime français, 1635-1760*. Québec, 1911.

Harris, R. S. and I Montagnes, eds. *Cold Iron and Lady Godiva: Engineering Education at Toronto, 1920-72*. Toronto, 1973.

Harris, R. S. *A History of Higher Education in Canada, 1663-1960*. Toronto, 1976.

Jarrell, R. A. "Science Education at the University
of New Brunswick in the Nineteenth Century,"
Acadiensis (Spring, 1973), 55-79.

Nicholls, R. V. V. "A Century and a Quarter of
Chemistry at McGill University," *Canadian
Chemical and Process Industries* 28 (1944),
559-66, 580.

Smith, E. C. *Department of Biology, Acadia
University, 1910-1960*. Kentville, 1961.

"Twenty-five years of Chemistry at Laval," *Canadian
Chemical and Process Industries* 29 (1945),
357-64.

Young, C. R. *Early Engineering Education at Toronto,
1851-1919*. Toronto, 1958.

F. SCIENTIFIC AND TECHNICAL INSTITUTIONS/INSTITUTIONS
SCIENTIFIQUES ET TECHNIQUES

Alcock, F. J. *A Century in the History of the
Geological Survey of Canada*. Ottawa, 1947.

Bowler, P. J. "The Early History of Scientific
Societies in Canada," in A. Oleson and S.
Brown, *The Pursuit of Knowledge in the Early
American Republic*. Baltimore, 1976, pp. 326-39.

Canada Agriculture: the First Hundred Years.
Ottawa, 1967.

Canada Patent Office. *Patents of Canada, 1824-1855*.
Toronto, 1865.

Canada Patent and Copyright Office. *The Canadian
Patent Office Records*. Ottawa, 1873-1969.

*Centenary Volume of the Literary and Historical
Society of Quebec*. Quebec, 1924.

Denison, M. *The Barley and the Stream: the Molson
Story*. Toronto, 1955.

Denison, M. *Harvest Triumphant: the Hundred Year
Story of Massey-Harris*. Toronto, 1948.

Eggleston, W. *National Research in Canada: The NRC,
1916-66*. Toronto, 1978.

Fergusson, C. B. "Mechanics' Institutes in Nova
Scotia," *Bulletin of Public Archives of Nova
Scotia* 14 (1960).

Goodspeed, O. J. *DRB: A History of the Defence
Research Board*. Ottawa, 1958.

Johnstone, K. *The Aquatic Explorers: A History of the Fisheries Research Board of Canada.* Toronto, 1977.

Thistle, M. *The Inner Ring.* Toronto, 1965.

Thistle, M. *The Mackenzie-MacNaughton Wartime Letters.* Toronto, 1975.

Wallace, W. S. *The Royal Canadian Institute Centenary Volume.* Toronto, 1949.

Zaslow, M. *Reading the Rocks: A History of the Geological Survey of Canada, 1842-1972.* Toronto, 1975.

G.   BIOGRAPHICAL STUDIES/ETUDES BIOGRAPHIQUES

Adams, E. D., "Biographical Memoir of Thomas Sterry Hunt," *Biographical Memoirs of the National Academy of Science 15* (1934), 207-38.

Bailey, J. W. *Loring Woart Bailey: The Story of a Man of Science.* Saint John, 1925.

Barkhouse, J. C. *George Dawson, the Little Giant.* Toronto, 1974.

Bassett, J. and A. R. Petrie. *William Hamilton Merritt: Canada's Father of Transportation.* Toronto, 1974.

Burpee, L. J. *Sandford Fleming, Empire Builder.* London, 1915.

Corbett, E. A. *Henry Marshall Tory, Beloved Canadian.* Toronto, 1954.

Cushing, H. *The Life of Sir William Osler.* Oxford, 1925.

Dawson, J. W. *Fifty Years of Scientific and Educational Work in Canada.* Montreal, 1901.

Gillen, M. *The Masseys: Founding Family.* Toronto, 1966.

Greening, W. E. *Sir Casimir Stanislas Gzowski.* Toronto, 1959.

Harrington, B. J. *The Life of Sir William Logan.* Montreal, 1883.

Huard, V. A. *La vie et l'oeuvre de l'abbé Provancher.* Québec, 1926.

Lamontagne, R. *Roland-Michel Barrin de la Galissonière 1693-1756.* Québec, 1970.

Langton, H. H. *Sir John Cunningham McLennan.*
Toronto, 1939.

Loudon, W. J. *A Canadian Geologist.* Toronto, 1930.
--on J. B. Tyrrell

MacLean, H. *Man of Steel: The Story of Sir Sandford
Fleming.* Toronto, 1969.

Macoun, J. *Autobiography.* Ottawa, 1922.

Mayles, S. *William van Horne.* Toronto, 1975.

Murphy, L. J. *Thomas Coltrin Keefer.* Toronto, 1976.

O'Brien, C. F. *Sir William Dawson. A Life in Science
and Religion.* Philadelphia, 1971.

Pomeroy, B. M. *William Saunders and his Five Sons.*
Toronto, 1956.

Rousseau, J. and W. G. Dore, "L'Oublié de l'histoire
de la science canadienne-George Lawson, 1827-
1895," in G. F. G. Stanley, ed., *Pioneers of
Canadian Science.* Toronto, 1966, pp. 54-80.

Russell, L. S. "'Carbide' Willson," *Canada,* vol. 3,
no. 1 (September, 1975), pp. 20-33.

Sheffe, N. *Casimir Gzowski.* Toronto, 1975.

Vallée, A. *Michel Sarrazin.* Québec, 1927.

Walker, F. N. *Daylight through the Mountains:
Letters and Labours of Civil Engineers Walter
and Francis Shanly.* Toronto, 1959.

Winslow-Spragge, L. *Life and Letters of George Mercer
Dawson.* Montreal, 1962.

H. HISTORY OF MEDICINE/HISTOIRE DE LA MEDECINE

Abbott, M. *A History of Medicine in the Province of
Quebec.* Montreal, 1931.

Agnew, G. H. *Canadian Hospitals, 1920-1970: A
Dramatic Half-Century.* Toronto, 1975.

Angus, M. *Kingston General Hospital: A Social and
Institutional History.* Montreal, 1973.

Birkett, H. S. *A Brief Account of the History of
Medicine in the Province of Quebec from 1535
to 1838.* New York, 1908.

Bull, W. P. *From Medicine Man to Medical Man.*
Toronto, 1934.

Canniff, W. *The Medical Profession in Upper Canada, 1783-1859.* Toronto, 1894.

Cosbie, W. G. *The Toronto General Hospital 1819-1965: A Chronicle.* Toronto, 1975.

Gibbon, J. and M. Mathewson. *Three Centuries of Canadian Nursing.* Toronto, 1947.

Godfrey, C. M. *The Cholera Epidemics in Upper Canada, 1832-1866.* Toronto, 1968.

Gullett, D. W. *A History of Dentistry in Canada.* Toronto, 1971.

Hacker, C. *The Indomitable Lady Doctors.* Toronto, 1974.

Heagerthy, J. J. *Four Centuries of Medical History In Canada.* 2 volumes. Toronto, 1928.

Howell, W. B. *Medicine in Canada.* New York, 1933.

Jackson, H. M. *The Story of the Royal Canadian Dental Corps.* n.p., 1956.

MacDermot, H. E. *History of the Canadian Medical Association, 1867-1921.* Toronto, 1935.

MacDermot, H. E. *One Hundred Years of Medicine in Canada, 1867-1967.* Toronto, 1967.

Mignault, L.-D. "Histoire de l'école de médecine et de chirurgie de Montréal," *Union Médical du Canada* 55 (1926), 598.

Morin, V. "L'évolution de la médecine au Canada français," *Cahiers des Dix,* 25 (1960), 65-83.

Nicholson, G. W. L. *Seventy Years of Service: A History of the Royal Canadian Army Medical Corps.* Montreal, 1976.

Pariseau, L. *Le centenaire de la fondation du "Journal de Médecine de Québec."* Montréal, 1926.

Rose, T. F. *From Shaman to Modern Medicine: A Century of the Healing Arts in British Columbia.* Victoria, 1972.

I. SCIENCE, TECHNOLOGY, AND SOCIETY/LES SCIENCES, LA TECHNOLOGIE ET LA SOCIETE

Babbitt, J. D., ed. *Science in Canada: Selections from the Speeches of E. W. R. Steacie.* Toronto, 1965.

231

Bradwin, E. *The Bunkhouse Man: A Study of Work and Play in the Camps of Canada, 1903-1914.* Toronto, 1972.

C.I.S.T.I. *Scientific Policy, Research and Development in Canada: Bibliography.* (I for 1935-70; II for 1970-72; III for 1972-75; Supplement for 1975-77.)

Craig, G. M. *Early Travellers in the Canadas 1791-1867.* Toronto, 1955.

Cross, M. *The Workingman in the Nineteenth Century.* Toronto, 1974.

Doern, G. B. *Science and Politics in Canada.* Montreal, 1972.

Duchesne, R. *La science et le pouvoir au Québec, 1920-1965.* Québec, 1978.

Grant, G. *Technology and Empire. Perspectives on North America.* Toronto, 1969.

Hayes, R. *The Chaining of Prometheus.* Toronto, 1971.

Jarrell, R. A. "The Rise and Decline of Science at Quebec: 1824-44," *Social History/Histoire sociale* 9:1 (1977), 77-91.

Senate Special Committee on Science Policy, *A Science Policy for Canada* (Lamontagne Report). 3 vols. Ottawa, 1971-73.

Young, C., H. Innis and J. Dales. *Engineering and Society with Special Reference to Canada.* Toronto, 1947.

APPENDIX D

APPENDICE D

CONFERENCE PARTICIPANTS

PARTICIPANTS AU CONGRES

John W. Abrams, Professor
of Industrial Engineering,
University of Toronto.

Norman R. Ball, Science
and Engineering Archivist,
Public Archives of Canada.

Jacques Bernier, Professeur
adjoint d'histoire,
Université Laval.

Peter J. Bowler, Assistant
Professor of History,
University of Winnipeg.

Louise Dandurand, Chargé de
cours, Sciences politiques,
Université de Ottawa.

Raymond Duchesne, Institut
d'histoire et sociopolitique
des science, Université de
Montréal.

Margaret Dunn, Medical Ar-
chives Project, Hannah
Institute for History of
Medicine.

Phillip C. Enros, Visiting
Lecturer, Institute for
History and Philosophy of
Science and Technology,
University of Toronto.

John Farley, Associate
Professor of Biology,
Dalhousie University.

C. E. S. Franks, Associate
Professor of Political
Studies, Queen's University

Toby Gelfand, Associate
Professor and Hannah Chair
for History of Medicine,
University of Ottawa.

Peter Gillis, Manpower and
Social Development Archivist,
Public Archives of Canada

J. W. Grove, Professor of
Political Studies,
Queen's University.

Jean-Claude Guédon, Pro-
fesseur agrégé, Institut
d'histoire et sociopoli-
tique des sciences,
Université de Montréal.

Sandra Guillaume, Archi-
vist, Ontario Multicultur-
al History Society,
Toronto, Ontario.

Richard A. Jarrell, Assoc-
iate Professor of Natural
Science, Atkinson College,
York University.

Trevor H. Levere, Assoc-
iate Professor, Institute
for History and Philosophy
of Science and Technology,
University of Toronto.

Ian MacPherson, Associate
Professor of History,
University of Victoria.

Wendy Mitchinson, Assis-
tant Professor of History,
University of Windsor.

Ian Montagnes, General Editor, University of Tor-
onto Press.

Yves Mougeot, Research
Grants officer, Social Science and Humanities Re-
search Council of Canada.

David Newlands, Coordina-
tor, Museology Programme,
University of Toronto.

William Ormsby, Archivist
of Ontario, Toronto, Ont-
ario.

Robert W. Passfield, Nat-
ional Historic Parks and
Sites, Indian and Northern
Affairs Department,
Ottawa.

G. R. Paterson, Executive
Director, Hannah Foundation
for the History of Medicine,
Toronto, Ontario.

James O. Petersen,
Research Associate,
Institute for History
and Philosophy of Science
and Technology, University of Toronto.

Christopher J. B. L. Porter,
Curator of Technology,
Upper Canada Village,
Morrisburg, Ontario.

Arthur J. Ray, Associate
Professor of Geography,
York University.

R. A. Richardson, Associate
Professor of History of
Medicine and Science,
University of Western
Ontario.

W. George Richardson,
Associate Professor of
Engineering,
Queen's University.

Charles G. Roland, Hannah
Professor of History of
Medicine, McMaster
University.

Arnold E. Roos, National
Historic Parks and Sites,
Indian and Northern Affairs
Department, Ottawa.

David Rudkin, University
Archivist, University of
Toronto.

Loris S. Russell, Curator
and Professor Emeritus,
Royal Ontario Museum and
University of Toronto.

Brigitte Schroeder-Gudehus,
Professeur titulaire, In-
stitut d'histoire et sociopolitique des sciences,
Université de Montréal.

John W. Scott, Professor of
Medicine, University of
Toronto.

George Sinclair, President,
Sinclair Radio Laboratories,
Ltd., Concord, Ontario.

J. Bruce Sinclair, Director,
Institute for History and
Philosophy of Science and
Technology, University of
Toronto.

J. W. T. Spinks, President
Emeritus, University of
Saskatchewan.

Philip Teigen, Librarian,
Osler Library, McGill
University.

A. W. Tickner, Senior Archival Officer, National
Research Council, Ottawa.

R. J. Uffen, Dean of Applied
Science, Queen's University.

Vittorio de Vecchi, History
and Social Studies of Science Unit, University of
Sussex.

Roger Voyer, Director of
Research, Science Council
of Canada, Ottawa.

David F. Walker, Associate
Professor of Geography,
University of Waterloo.

James Wardrop, Associate
Curator of History, Brit-
ish Columbia Provincial
Museum, Victoria.

Hugh R. Wynne-Edwards, Assistant Secretary, Ministry
of State for Science and
Technology, Ottawa.

INDEX

INDEX OF NAMES AND INSTITUTIONS

INDEX DES NOMS ET DES INSTITUTIONS

(Text and Appendix A/Texte et appendice A)

246